T0349012

Engineered Nanopores for Bioanalytical Applications

Engineered Nanopores for Bioanalytical Applications

Joshua B. Edel

Tim Albrecht

AMSTERDAM • BOSTON • HEIDELBERG • LONDON
NEW YORK • OXFORD • PARIS • SAN DIEGO
SAN FRANCISCO • SINGAPORE • SYDNEY • TOKYO
William Andrew is an imprint of Elsevier

William Andrew is an imprint of Elsevier
225 Wyman Street, Waltham, MA 02451, USA
The Boulevard, Langford Lane, Kidlington, Oxford OX5 1GB, UK

Notice
Knowledge and best practice in this field are constantly changing. As new research and
experience broaden our understanding, changes in research methods or professional practices, or medical treatment may become necessary.

Practitioners and researchers must always rely on their own experience and knowledge in
evaluating and using any information, methods, compounds, or experiments described
herein. In using such information or methods they should be mindful of their own safety
and the safety of others, including parties for whom they have a professional responsibility.

To the fullest extent of the law, neither the Publisher nor the authors, contributors, or
editors, assume any liability for any injury and/or damage to persons or property as a
matter of products liability, negligence or otherwise, or from any use or operation of any
methods, products, instructions, or ideas contained in the material herein.

British Library Cataloguing-in-Publication Data
A catalogue record for this book is available from the British Library

Library of Congress Cataloging-in-Publication Data
A catalog record for this book is available from the Library of Congress

ISBN: 978-1-4377-3473-7

For information on all Elsevier publications
visit our website at elsevier direct.com

Printed and bound by CPI Group (UK) Ltd, Croydon, CR0 4YY

Working together
to grow libraries in
developing countries

www.elsevier.com • www.bookaid.org

List of Contributors

T. Albrecht
Department of Chemistry, Imperial College London, London

Mariam Ayub
Department of Chemistry, University of Oxford, United Kingdom

Kevin J. Freedman
Department of Chemical and Biological Engineering, Drexel University, Philadelphia, PA

Thomas Gibb
Department of Chemistry, Imperial College London, London

Gaurav Goyal
School of Biomedical Engineering, Science and Health Systems, Drexel University, Philadelphia, PA

Min Jun Kim
School of Biomedical Engineering, Science and Health Systems, Drexel University, Philadelphia, PA; Department of Mechanical Engineering and Mechanics, Drexel University, Philadelphia, PA

Ulrich F. Keyser
Cavendish Laboratory, University of Cambridge, Cambridge, UK

P. Nuttall
Department of Chemistry, Imperial College London, London

Oliver Otto
Biotechnology Center, Technical University Dresden, Tatzberg Dresden, Germany

Anmiv S. Prabhu
School of Biomedical Engineering, Science and Health Systems, Drexel University, Philadelphia, PA

Joseph W.F. Robertson
Semiconductor and Dimensional Metrology Division, National Institute of Standards and Technology, Gaithersburg, MD, USA

Vincent Tabard-Cossa
Center for Interdisciplinary NanoPhysics, Department of Physics, University of Ottawa, 150 Louis Pasteur, Ottawa, ON, K1N 6N5, Canada

Contents

Introduction

Engineered nanopores for bioanalytical applications

Nanopore sensor devices emerge as a new, powerful class of single-molecule sensors that allow for the label-free detection of biomolecules (DNA, RNA, and proteins), non-biological polymers, as well as small molecules. In the context of this book, a nanopore sensor specifically refers to devices that consist of a thin, highly insulating membrane, with either a single or array of well-defined nanometre-sized channels ("nanopores"). Integrated into a liquid cell, the membrane separates the latter into two compartments—as a result transport of electrolyte solution, ions, and analyte molecules between the compartments can only occur through the nanopore(s). Most commonly, transport is driven by the application of an electric field, inducing an ionic current that is determined by the cross-sectional area and length of the pore channel. The passage of individual analyte molecules through the pore modulates the pore conductance, and thus the electric current through the cell. Such a "translocation event" constitutes the primary sensing signal. That said, device operation is by no means restricted to ionic current detection. For example, techniques such as force microscopy, optical detection, and tunneling spectroscopy have all been gaining in popularity. These have been particularly interesting prospects for large-scale parallel operation (multiplexing) or biomolecular analysis with ultra-high spatial resolution, for example, for DNA sequencing. Naturally, every sensing modality has both advantages and disadvantages; however, it is clear that the nanopore platform allows for great flexibility when it comes to the choice of detection strategies. One exciting prospect that has only recently started to be explored is in combining the detection strategies to maximize the information that can be extracted from probing a single molecule.

Broadly speaking, nanopore sensor devices fall into two categories, namely "biological" and "solid-state." Biological nanopores are made from a pore-forming protein or an ion channel that typically self-assembles into a thin, highly insulating lipid bilayer membrane. Pore "fabrication" then relies on biochemistry, namely protein expression and modification, purification, and ultimately self-assembly into the membrane. Since the protein structure is dictated by the amino acid composition, biological pores can easily be made in large quantities (at least by single-molecule standards) with extremely high reproducibility of the channel dimensions and properties. Accordingly, the electric current through the pore is very stable with an excellent signal-to-noise ratio. On the other hand, while the protein pores themselves withstand rather drastic physical conditions (e.g., temperatures up to the boiling point of water), the lipid bilayer membrane tends to be the "weak link" in this type of sensor, in terms of its mechanical, thermal, and electric stability. This is particularly critical for real-life operations outside the research laboratory. In addition, the high reproducibility of the pore dimensions comes at the price of reduced flexibility. In order to achieve maximum signal-to-noise ratio and to avoid clogging, the pore diameter should be slightly larger than the molecular dimensions of the analytes of interest (note: the cross-sectional diameter of double-stranded DNA is approximately 2.5 nm). If the dimensions do not match for a particular application, the best case scenario is simply to choose a different protein containing either a larger or smaller pore. In the worst case, no such pores may be available or potentially not even exist, prohibiting their use all together. Perhaps one of the most popular proteins is α-hemolysine (αHL). αHL has been extensively used in the literature and has primarily found applications in the detection of small molecules or single-stranded DNA. Double-stranded DNA is already too large to fit through the 1.4 nm αHL pore without structural alterations.

In "solid-state" pores the lipid bilayer is replaced with a synthetic membrane, usually made of silicon nitride or silicon dioxide, and also more recently with graphene. The pore opening is then drilled with a focused ion- or electron-beam, or defined lithographically. While solid-state nanopore devices tend to show worse electric performance, in terms of signal-to-noise ratio, and rather less well-defined surface chemistry than biological pores (i.e., the composition of biological pores such as αHL are known with atomic precision as such they have a well-defined crystal structure), they feature a greater flexibility in the pore dimensions, are mechanically robust, and the surfaces can be easily modified (e.g., silane chemistry). Furthermore, the devices tend to be fabricated using well-developed sophisticated semiconductor processing technology resulting in significant flexibility when incorporating electronics.

Clearly, choices need to be made when selecting the material for the pore. Unfortunately, it is not as simple as suggesting "one shoe will fit all" but rather the pore material has to be carefully selected in order to maximize the capabilities of the platform while at the same time ensuring analyte compatibility. Regardless of whether a solid-state or biological pore is used the flexibility of the platform ensures that nanopore sensing can and has been opened up to a plethora of other target analytes aside from DNA, including proteins, protein/protein or protein/DNA complexes, and applications beyond sequencing, for example, in molecular fingerprinting or genetic profiling.

This brief outline already indicates an important aspect of nanopore research, namely that it is a highly interdisciplinary field. Its foundation rests on biochemistry, condensed-matter and single-molecule biophysics, electrochemistry and surface science, molecular self-assembly, electronics engineering, materials science, and state-of-the-art statistical analysis. Nanopore-based sensing provides insight into the fundamentals of transport processes in confined environments, sheds light on single-molecule structure and dynamics, and pushes the boundaries of the underlying sensing technologies. The resulting applications are equally diverse and range from membrane science to the detection of DNA mutation and damage.

For newcomers as well as experienced players in the field, this interdisciplinarity is a challenge, in that mastery of all these areas is a prerequisite to success. What may be trivial to a chemist might be highly challenging to an electronics engineer and vice versa—so collaboration is key. It also puts particular focus on how to transmit the required basic knowledge and to educate future students in the field, and facilitate collaboration between the different disciplines. Although the majority of the material presented in this book is likely to be found in the respective specialized literature, it can often be rather difficult to find; especially for someone new to the field or for someone coming from a different background. As editors, we hope that this book and the references provided will guide the reader toward these specialized texts, when needed. In some cases, the authors also share their "tricks of the trade," which should be of interest for anyone working in this area. As such this book is intended for those who are new to the field as well as the seasoned expert wanting a little more insight into nanopore sensing. Accordingly, this book is broken down into six chapters. The first two chapters cover the fundamentals of ion transport and the biophysics of the DNA translocation process. Chapter 3 develops the discussion on noise and low current electronics giving specific insight on how best to optimize the nanopore platform. This is followed by Chapter 4 on case studies using biological pores. The final two chapters relate to fabrication strategies of solid-state nanopore sensors and respective case studies. We would like to take the opportunity to express our gratitude and thanks to all the contributors who took up our offer of writing a chapter.

It is clear that a book on this subject is never "complete," especially when considering the dynamic nature of the field. We estimate that more than 100 research groups are active in this area worldwide, both in academia and increasingly also in industry. To this end, it is worth noting that commercial, nanopore-based sensor devices are beginning to enter the market place, particularly in high-speed DNA sequencing. These developments may raise the profile of nanopore sensing even more, making a book such as this one ever more relevant.

We hope that the reader finds the content interesting and relevant to their own endeavors, but would appreciate any feedback and comments on the content, its presentation, or the wider context.

Tim Albrecht and Joshua B. Edel
Department of Chemistry
Imperial College, London
November 2012

Ion Transport in Nanopores

T. Albrecht, T. Gibb and P. Nuttall

Department of Chemistry, Imperial College London, London

CHAPTER OUTLINE

1.1 Introduction

The transport of ions and other charged species, as well as liquid, is at the heart of any nanopore sensor, Figure 1.1. The flux of ions determines the electric current, which is often used as the sensing signal, but it also affects the liquid, for example in electroosmosis. Ion and liquid motion may also have an impact on the translocation of biopolymers through the nanopore, such as DNA or proteins, both in terms of the duration and magnitude of the concomitant ion current modulation.

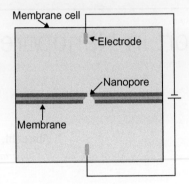

FIGURE 1.1

Schematic of a nanopore sensor. The membrane separates the two liquid compartments of the cell with a single (or small number of) nanopore(s) allowing the exchange of ions and liquid. A bias voltage between two electrodes—one on each side of the membrane—allows electric field-driven transport through the nanopore. The current sensing electronics are not shown.

The fundamental physics of ion motion has been long established and forms the basis for the theoretical understanding of charge transport in semiconductor physics, electrochemistry, membrane science, and other areas. This chapter does not attempt to cover the vast area of electrokinetic fluid and ion transport but rather give a brief introduction into the main factors governing ion transport in nanopores with a focus on mean-field, dielectric continuum theory, and the Nernst–Planck equation. References to the more advanced literature are given where appropriate.

Simple models have the advantage that the resulting analytical expressions for the pore conductance G_{pore} and other parameters are transparent and potentially yield direct physical insight. Widely used "work relations" emerge from a relatively simple theory, which can be used to validate experimental data, for example, when assessing the pore dimensions. The lack of complexity, on the other hand, also implies certain limitations, in terms of quantitative accuracy and the validity range. For very small nanopores, in particular, where the pore size reaches "molecular" dimensions, the continuum model is expected to break down. Continuum theories have, however, an advantage compared to more sophisticated models and computer simulations in that they allow the entire electrochemical cell—including electrodes, electrolyte solution, and the membrane—to be treated with little extra difficulty. For example, in a conventional two-electrode system, the electric current is determined by the potential distribution in the entire cell, including potential drops at electrode/solution interfaces, the solution itself, and across the pore. If the pore is small and long, its resistance will usually dominate the total cell resistance and focusing on the pore itself is sufficient to capture all relevant features of the nanopore sensor. On the other hand, if the pore channel is very short—such as in graphene-based nanopores—the potential gradient between the electrodes and the pore openings is important and the so-called "access resistance" determines the overall pore resistance. This effect has been addressed already by Hille and Hall in the 1970s, who derive simple analytical expressions [1−3]. Finally, nanopore devices with multiple electrodes as potential current sources can readily be treated, as shown by Albrecht. However, careful consideration needs to be given to the limitations of the model used with regard to potential over-parameterization for example [4].

The Nernst−Planck, the Poisson, and the Navier−Stokes equations are known as the governing equations for the ion flux, local charge distribution, and fluid dynamics (momentum transport), respectively. At a higher level of complexity and accuracy, these are solved self-consistently for a particular geometry. The results are remarkably accurate but "only" numerical. Molecular-level detail may be included in Brownian or Langevin dynamics simulations at increasing computational expense. Detailed reviews on these more advanced topics may be found in Refs [5−7].

In order to understand the working principle, potential, and limitations of a nanopore sensor, we first need to look at the fundamental physical processes that underlie ionic and molecular motion in solution. Brownian motion is present at all times in a nanopore sensing experiment and offers a useful reference point to assess relevant length and timescales, depending on the species' diffusion coefficient D. We will then move on to discuss ion transport in the presence of an external driving force, such as a concentration gradient (diffusion) or an electric field (migration). Based on the Langevin model, we derive the Nernst−Planck equation for the flux of ions and an expression for the steady-state ion current for different pore geometries, taking into account surface charge effects where appropriate. Finally, we describe a basic model for particle translocation through nanopores after deBlois and Bean [8].

1.2 Brownian motion

Brownian motion is the random motion of particles suspended in a fluid due to continual collisions with fast moving solvent molecules. With large numbers of molecules hitting the Brownian particle at any one time, momentum is transferred and the particle moves in a random, uncorrelated fashion. The mean square displacement in one dimension is given by:

$$\overline{x^2} = 2Dt \tag{1.1}$$

where x is the displacement, D the diffusion coefficient, and t the time. Examining the equation reveals that it predicts substantial timescales for any macroscopic displacements but also indicates that Brownian motion can be considered a rapid process at the nanometer scale. Using $D = 10^{-10}$ m^2/s as typical of a small protein and the one-dimensional model above, the particle would take roughly 13 µs to move along a pore channel of 50 nm length. This value would constitute a speed limit to translocation, since surface interactions and other such effects are ignored, and should only be taken as a crude approximation. More accurately, the escape time for a Brownian particle is described by the "narrow escape model" [9,10].

Brownian motion is always present in a fluid, even in the presence of strong external fields, and causes some degree of spatial and electronic noise. Due to Brownian motion, one would expect translocating particles to have a velocity and thus translocation time distribution, even though other factors such as particle–wall interactions, fluid dynamics, and other effects may be more significant in magnitude. The potential for a particle to move backward or forward through a nanopore during translocation also causes complications in nanopore sensing and in particular nanopore-based DNA sequencing. To this end, Brownian motion enhances the error rate, for example, by causing particular bases to be read twice or not at all [11].

1.3 Net transport of ions: the Nernst–Planck equation and its derivation

The derivation of the Nernst–Planck equation for ionic flux starts with Langevin's analysis of random motion [5]. If an individual charged particle is immersed in a molecular solvent at a given constant temperature T, it is subjected to random collisions (Brownian motion), which change the particle momentum and result in an instantaneous velocity of the particle $v = dx/dt$. The fluctuating force on particle can be considered as two individual forces: the first, a random force $\mathbf{F}_r(t)$, is likely to average to zero over a number of collisions; while the second, a velocity-dependent drag force $\mathbf{F}_{drag}$, will exist as a slow fluctuation. In the case of an applied electric field, an additional force can be written as $\mathbf{F}_e = z \cdot e \cdot E$, which acts on a particle with nonzero charge z. Newton's second law is then:

$$m \cdot \frac{d\mathbf{v}}{dt} = \mathbf{F}_e + \mathbf{F}_{drag} + \mathbf{F}_r(t) \tag{1.2}$$

The drag force $\mathbf{F}_{drag}$ is a function of the drift velocity $\mathbf{v}_{drift}$, which may be expanded in a power series. Neglecting all higher-order terms, $\mathbf{F}_{drag}$ is then approximately given by $-\alpha\,\mathbf{v}_{drift}$, where α is a proportionality factor.

Thus, the instantaneous velocity is $\mathbf{v} = \mathbf{v}_{drift} + \mathbf{v}_r$, where $\mathbf{v}_r$ averages to zero after a short time. Equation (1.2) then becomes:

$$m \cdot \left(\frac{d\mathbf{v}_{drift}}{dt} + \frac{d\mathbf{v}_r}{dt} \right) = z \cdot e \cdot E - \alpha \cdot \mathbf{v}_{drift} + \mathbf{F}_r(t) \tag{1.3}$$

where both dv_r/dt and $\mathbf{F}_r$ average to zero after a short time. At steady-state, when $\mathbf{v}_{\text{drift}}$ is independent of time, one obtains for the drift velocity:

$$\mathbf{v}_{\text{drift}} = \frac{1}{\alpha} \cdot z \cdot e \cdot E = u_i \cdot \textbf{force} \tag{1.4}$$

Here, u_i is the electrophoretic mobility, $u_i = 1/\alpha$, and E is the electric field. The general result in Eq. (1.4) can include diffusion and electric field contributions toward the net velocity of $\mathbf{v}_{\text{drift}}$. Note that we chose to restrict ourselves to an external electric force in the derivation above.

If the solvent can be regarded as static and we focus on ion transport only, the flux of an ionic species i is then simply the number density (or concentration) times $\mathbf{v}_{\text{drift}}$.

$$j_i = c_i \cdot \mathbf{v}_{\text{drift},i} = c_i \cdot u_i \cdot \textbf{force} = -c_i \cdot u_i \cdot \frac{d\mu_{\text{el}}}{dx} \tag{1.5}$$

μ_{el} is the electrochemical potential and the force is proportional to its gradient in the direction of transport. At a given temperature μ_{el} is dependent on the activity of the ion $a_i = \gamma_i \cdot c_i$, where γ_i is the activity coefficient. In molar units, one obtains:

$$\mu_{\text{el}} = \mu^0 + RT \cdot \ln c_i + RT \cdot \ln \gamma_i + z_i \cdot F \cdot \Phi + p \cdot V_m \tag{1.6}$$

μ^0 is the standard chemical potential, R and T the universal gas constant and temperature, Φ the local electrostatic potential, and V_m the molar volume. z_i is the ionic charge, p pressure and F Faraday's constant.

In dilute solution, the electrophoretic mobility u_i is related to the diffusion coefficient D_i through the Einstein relation: $u_i = D_i/RT$.[1]

Substituting Eq. (1.6) into Eq. (1.5), one obtains in the absence of any pressure gradients:

$$
\begin{aligned}
j_i &= -c_i \cdot \frac{D_i}{RT} \cdot \frac{d}{dx} \left[\mu^0 + RT \cdot \ln c_i + RT \cdot \ln \gamma_i + z_i \cdot F \cdot \Phi \right] \\
&= -D_i \cdot \left[c_i \cdot \frac{d \ln c_i}{dx} + c_i \cdot \frac{d \ln \gamma_i}{dx} + c_i \cdot \frac{z_i \cdot F}{RT} \cdot \frac{d\Phi}{dx} \right] \\
&= -D_i \cdot \left[\frac{dc_i}{dx} + c_i \cdot \frac{z_i \cdot F}{RT} \cdot \frac{d\Phi}{dx} \right] \\
&= -D_i \cdot \left[\frac{dc_i}{dx} - c_i \cdot \frac{z_i \cdot F}{RT} \cdot E \right]
\end{aligned} \tag{1.7}
$$

---

[1]Note that the "absolute" mobility u_i (formally in units of speed/unit force) needs to be distinguished from the so-called "conventional" mobility u_{conv}, which is the drift velocity in 1 V/m electric field and hence in units of m^2/Vs. Both are related via $u_{\text{conv}} = u_i \cdot z_i \cdot e$. Since the electric force on an ion is $z_i \cdot e$ times the electric field E, $F = z_i \cdot e \cdot E$, it is u_{conv} that relates the drift speed v_d to E, $v_d = u_{\text{conv}} \cdot E$.

The activity coefficient was taken to be independent of distance (or location). From Eq. (1.7), it is apparent that the total flux of ion i is governed by diffusion (first term, Fick's first law for $E = 0$) and migration (second term).

In the absence of space charge effects, the electric current density J is simply the sum of all fluxes for each ion in solution:

$$J = e \cdot \sum_i z_i \cdot j_i \qquad (1.8)$$

If j_i is in molar units, e needs to be replaced by Faraday's constant F. Equations (1.7) and (1.8) can serve as the basis for deriving useful expressions for the pore conductance. Note, however, that the above Coulombic model is over-simplified in many ways. For example, as written it includes the formal ionic charge z_i, rather than the effective charge, which is concentration dependent [12]. Electrostatic screening of the ionic charge by the solvation shell[2] and dynamic effects that originate from the ion moving in an external field such as an electric or chemical potential gradient are ignored.

In the absence of an external electric field, the solvation shell can be regarded as symmetric. However, upon application of an electric field, the ion moves with an effective drift velocity v_{drift} along the potential gradient. This results in a distortion of the ionic cloud surrounding the ion, where the solvation cloud builds up in front of the moving ion and decays behind it. This produces an additional electric field, the so-called "relaxation field," which counteracts the external field and therefore slows ion motion in solution. It should be noted, however, that this effect can be neglected for very small or large ratios of radius of curvature to double layer thickness, i.e., $r/\lambda_D < 0.1$ or > 300 where λ_D is in the order of 0.3 nm for 1 M KCl buffer (discussed later in the chapter) [13].

Secondly, the external electric field not only acts on the central ion, but also on the oppositely charged solvation shell. Both entities thus have a tendency to move in opposite directions, giving rise to an electrophoretic force, which again counteracts the external driving field.

Both relaxation and electrophoretic corrections may be incorporated into the expression for the mobility and thus the flux [14]. The magnitude of these corrections depend on the characteristics of the solvation sphere and the properties of the central ion. The inverse Debye screening length λ_D^{-1} can be used to include these effects in the flux:

$$\lambda_D^{-1} = \sqrt{\frac{F^2 \cdot \sum c_i \cdot z_i^2}{\varepsilon RT}} \qquad (1.9)$$

$\varepsilon = \varepsilon_r \cdot \varepsilon_0$ is the permittivity of the solution and c_i the concentration of ion species i (in mol/m^3).

---

[2]For a detailed examination of solvation effects see Shaw (Chapter 2 "Ion–Solvent Interactions", p. 35) [13].

The concentration-dependent mobility is then:

$$u_i = u_{i,0} - \frac{1}{6\pi\eta}\lambda_{\rm D}^{-1} - u_{i,0}\frac{e\omega}{6\varepsilon z_i k_{\rm B}T}\lambda_{\rm D}^{-1} \qquad (1.10)$$

where the first term on the right-hand side, $u_{i,0}$, is the absolute mobility at infinite dilution, the second and third terms are the electrophoretic and relaxation correction, respectively. η is the dynamic viscosity of the medium (from Stokes' law) and ω is the constant factor that depends on the charge of the electrolyte ions ($\omega = 0.5859$ and 2.3436 for a 1:1 and 2:2 electrolyte, respectively). See Ref. [14] for derivations and further references.

Equation (1.7) may be rewritten to take into account both effects:

$$j_i = \left[u_{i,0} - \frac{1}{6\pi\eta}\lambda_{\rm D}^{-1} - u_{i,0}\frac{e\omega}{6\varepsilon z_i k_{\rm B}T}\lambda_{\rm D}^{-1}\right] \cdot c_i \cdot N_{\rm A} \cdot E \qquad (1.11)$$

$N_{\rm A}$ is Avogadro's number and ensures consistency of the units with regard to Eq. (1.6), i.e., if c_i is in mol/m^3.

It is clear from Eq. (1.10) that both of these correction effects increase in importance with increasing ionic strength, since the screening length $\lambda_{\rm D}$ becomes shorter. If $C_i \rightarrow 0$, $\lambda_{\rm D}^{-1} \rightarrow 0$ and the Debye length approaches infinity; in Eq. (1.10), $u \rightarrow u_{i,0}$ as expected.

While many nanopore experiments are performed at high salt conditions, say 0.1 M or even 1 M, where $\lambda_{\rm D} < 1$ nm for a 1:1 electrolyte, the investigation of surface charge effects typically includes variation of the ionic strength over several orders of magnitude. In this case, not only does the actual ion mobility affect the result, but also its concentration dependence; the flux of an ion is no longer simply linearly dependent on the concentration.

1.4 **The conductance of a pore with uncharged walls**

1.4.1 **Cylindrical pores**

In a typical nanopore sensing experiment, the electrolyte concentration is the same throughout the liquid cell. Ignoring non-ideal effects, the flux of ion i is according to Eq. (1.7):

$$j_i = D_i \cdot c_i \cdot \frac{z_i \cdot F}{RT} \cdot E \qquad (1.12)$$

For a 1:1 electrolyte such as KCl, Eq. (1.8) then yields the current density and finally an expression for the pore conductance ($c(K^+) = c(Cl^-) = c$).

$$J = \frac{z_i^2 \cdot F^2}{RT} \cdot [D_+ + D_-] \cdot c \cdot E = \sigma_{\rm s} \cdot E \qquad (1.13)$$

$\sigma_{\rm s}$ is the solution conductivity.

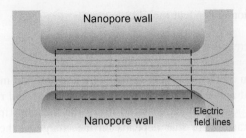

FIGURE 1.2

Cross-sectional view of a cylindrical pore with the membrane in the vertical direction. In long channels with small diameter, the electric field inside the pore is constant; edge effects on the pore conductance can be neglected.

If the length L of the nanopore channel is large and the diameter is small (assuming cylindrical geometry with cross-sectional area $A = \pi d^2/4$), Figure 1.2, the pore resistance R_{pore} dominates the total resistance of the cell. The potential drop across the pore $\Delta\Phi_{pore}$ is thus approximately equal to the applied voltage V_{bias}, provided that the potential drops at the electrode/solution interfaces $\Delta\Phi_{electr}$, and across the solution $\Delta\Phi_{sol}$ are small. Note that the latter two occur on both sides of the membrane and may or may not be split symmetrically between the two compartments. $\Delta\Phi_{electr}$ is small if (ideal) nonpolarizable electrodes such as Ag/AgCl electrodes with large electrochemically active areas are used. $\Delta\Phi_{sol}$ is small if the conductivity σ_s of the electrolyte is high. This is normally the case in nanopore translocation experiments, where KCl concentrations are in the range of $0.01-1$ M are normally used. Only if the pore channel is very short (very thin membranes), dose R_{pore} become small and the so-called "access resistance" dominant (*vide infra*).

$$V_{bias} = \Delta\Phi_{electr} + \Delta\Phi_{sol} + \Delta\Phi_{pore} \approx \Delta\Phi_{pore} \qquad (1.14)$$

Moreover, the electric field lines inside a long pore are approximately parallel and the electric field E is thus constant with $E \approx \Delta\Phi_{pore}/L \approx V_{bias}/L$. Hence, from Eq. (1.13) one obtains for the current I:

$$I = \frac{A}{L} \cdot \sigma_s \cdot V_{bias} = \frac{\pi d^2}{4L} \cdot \sigma_s \cdot V_{bias} \qquad (1.15)$$

Equation (1.15) is written such that positive V_{bias} produces formally positive currents. The pore conductance G_{pore} is then:

$$G_{pore} = \frac{dI}{dV_{bias}} = \frac{\pi d^2}{4L} \cdot \sigma_s = \frac{\pi d^2}{4L} \cdot \frac{z_i^2 \cdot F^2}{RT} \cdot [D_+ + D_-] \cdot c = \frac{\pi d^2}{4L} \cdot z_i^2 \cdot F^2 \cdot [u_+ + u_-] \cdot c$$

$$(1.16)$$

Note that u_+ and u_- are the absolute electrophoretic mobilities of the cation and anion, respectively; c is the concentration of ions in mol/m³, $c = c(K^+) = c(Cl^-)$.

As expected, the pore conductance depends on the properties of cation and anion. Both move in response to the external electric field, albeit in opposite directions. The current–voltage curve is then symmetric and no rectification occurs, i.e., the magnitude of the current does not depend on the sign of the bias voltage. Note, however, that Eq. (1.16) does not take into account space charge or electrophoretic and/or relaxation corrections. While Eq. (1.16) suggests a linear dependence of G_{pore} on c, nonlinearity enters in u_i and to some extent also z_i, as discussed above.

Spatial gradients in the activity coefficient γ_i were assumed to be absent, since the activity of the electrolyte on both sides of the nanoporous membrane was taken to be the same. This does not mean that the activity of the ions is necessarily the same inside the nanopore as in the bulk solvent. Depending on the ion distribution in the pore and the local electrostatic environment, they may well differ. For aqueous KCl solution, γ_i varies as a function of concentration from about 0.97 for 1 mM to 0.61 for 1 M KCl concentration [15]. This decrease in activity coefficient leads to the diffusion coefficient decreasing with total salt concentration, while the flux is also linearly related to the activity. The departure from ideal behavior of a solution therefore depends on the concentration of the electrolyte of interest.

1.4.2 **Pores with noncylindrical geometries**

So far, the pore geometry was assumed to be cylindrical with a high aspect ratio L/d. Depending on the fabrication process, nanopores are generally not perfectly cylindrical, but conical, hour glass-shaped or of a more complex geometry [16]. Such situations can be accounted for within the formalism derived above and the notion that the pore geometry may be represented by the sum of subsections or slabs. Each subsection possesses a differential resistance $dR(z)$, which are in series and thus add up to the total resistance R_{pore}. For a conical pore, each subsection is again cylindrical, but now with a position-dependent radius $r(z)$ and length dz (see Figure 1.3 for illustration).

$$dR(z) = \frac{1}{\sigma_s} \cdot \frac{dz}{\pi \cdot r(z)^2} \tag{1.17}$$

Integrating from $z = 0$ to $z = L$,[3] and introducing the opening angle α, so that $r(z) = r_{in} + z \tan(\alpha)$, one obtains [17,18]:

$$R_{pore}(\alpha) = \frac{1}{\sigma_s \pi} \cdot \frac{L}{r_{in} \cdot (r_{in} + L \cdot \tan(\alpha))} = \frac{1}{\sigma_s \pi} \cdot \frac{L}{r_{in} \cdot r_{out}} = \frac{4}{\sigma_s \pi} \cdot \frac{L}{d_{in} \cdot d_{out}} \tag{1.18}$$

If $d_{in} = d_{out} = d$ ($\alpha = 0$), Eq. (1.18) is simply the inverse of Eq. (1.16) for G_{pore}. An interesting implication of Eq. (1.18) is that R_{pore} is affected more by the small opening of pore, which can serve as a sensing area for translocation experiments, even if the total channel length is relatively large [19].

---

[3]The standard integral $\int (ax+b)^n dx = (ax+b)^{n+1}/a \cdot (n+1) + C$ may be useful in this context.

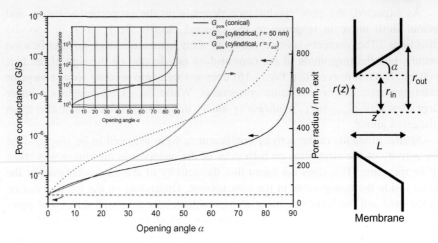

FIGURE 1.3

Left: G_{pore} as a function of opening angle α (left ordinate, solid line), according to Eq. (1.18); $r_{in} = 50$ nm, $L = 310$ nm; $c = 0.1$ M KCl. Dashed and dotted lines: conductance of cylindrical pores with $r = 50$ nm and $r = r_{out}(\alpha)$, respectively. Right ordinate: pore radius at the "exit" side, depending on α. Inset: $G_{pore}(\alpha)/G_{pore,cyl}$ ($r = 50$ nm) for different α. Right: illustration of the parameters used in Eqs (1.17) and (1.18).

Source: Figure reproduced with permission from Ayub et al. [17].

1.4.3 Access resistance

To simplify the discussion on pore conductance, edge effects are sometimes ignored, which is a reasonable assumption if the pore channel is long and narrow, $L \gg d$. For small L, this approximation breaks down and the so-called "access resistance" needs to be taken into account explicitly. Indeed, for short pores, the total resistance can be dominated by the access resistance.

The access resistance is the contribution of the electric field lines, converging from the bulk electrolyte to the entrance of the pore, Figure 1.4. Both Hille and Hall [1,2] have provided solutions for the access resistance to a pore; the treatments are identical to calculating the spreading resistance in metallic wires. The difference between the two solutions lies in the definition of the boundary between the pore and the bulk solution. Hille approximated the pore entrance as a hemisphere of the same radius as the pore [1]. The hemisphere contribution itself, towards the plane of the pore mouth, was assumed to be negligible. It is this contribution that is included in Hall's approach.

According to Hille, the access resistance is:

$$R_{access}^{Hille} = \frac{1}{\sigma_s \pi d} \tag{1.19}$$

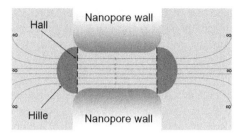

FIGURE 1.4

Cross-sectional view of a cylindrical pore with the membrane in the vertical direction now with focus on edge effects and illustration of the Hille and Hall models, respectively. If the channel is short, the access resistance is no longer small compared to the total resistance of the pore and has to be taken into account explicitly.

However, the hemisphere contribution is generally not small, as pointed out by Hall, who rederived the access resistance as the resistance between a hemispherical electrode at infinity and a flat disk electrode, representing the pore entrance [2].

In his derivation, Hall exploited the relation between the resistance between electrodes in conducting media and the capacitance between these electrodes in insulating media [20]:

$$R = \frac{\varepsilon}{\sigma_s C} \tag{1.20}$$

$\varepsilon = \varepsilon_r \cdot \varepsilon_0$ is the permittivity of the medium where the capacitance is measured. Calculating the capacitance of a conducting disk to a half spherical electrode at infinity gives:

$$C = 4\varepsilon d \tag{1.21}$$

The capacitance of a conducting disk on one side only is half of this total capacitance and thus the access resistance for one side of the nanopore is simply:

$$R = \frac{\varepsilon}{\sigma_s C} = \frac{\varepsilon}{2\sigma_s \varepsilon d} = \frac{1}{2\sigma_s d} \tag{1.22}$$

Since convergence effects occur at both ends of the pore, the total access resistance is twice of the value above (provided the system is symmetric):

$$R_{access}^{Hall} = \frac{1}{\sigma_s d} \tag{1.23}$$

Comparison of Eqs (1.19) and (1.23) shows that the Hall access resistance is a factor of $\pi/2$ larger than the access resistance according to Hille. Obviously, both are approximations to the real nanopore system, but there is good agreement between theory and experimental data [16,21].

The access resistance can be thought of as being in series with the resistance of the pore channel. Hence, the total resistance is the sum of both terms:

$$R_{\text{pore}} = \frac{1}{\sigma_s} \left(\frac{4L}{\pi d^2} + \frac{1}{d} \right) \tag{1.24}$$

The parameter dependence once again reiterates the points made above, namely that the relative contribution of the access term to R_{pore} increases for small L and small d. In the limit of an infinitely thin membrane, say similar to single-layer graphene [22,23], this expression reduces further to $G_{\text{pore}} = d \cdot \sigma_s$.

1.5 **The effect of surface charge**

1.5.1 **Charged surfaces in solution**

Surfaces immersed in a polar solution almost always carry a charge, due to the stabilizing effect of the solvent on ionic species. This may be caused by partial dissolution of the material, for example, in the case of metals where a small amount of metal ions enters the solution, or ionization of chemical functionalities on the surface, say deprotonation of —OH or —COOH groups or protonation of —NH₂ groups. The magnitude of the effect depends on the nature of the surface, the solvent and solution conditions such as ionic strength and pH, as well as the temperature. The principle applies equally to macroscopic surfaces of metals or insulators, colloids, nanoparticles, and biological species, including DNA, proteins, and other polyelectrolytes. Importantly, any mobile ions in the solution will be attracted to the charged surface for electrostatic reasons to compensate the charge. At the same time, thermal motion will tend to distribute ions homogeneously in the solution. In effect, the molecular structure at the solid/liquid interface can be rather complex, see Figure 1.5, and will be discussed later in this chapter.

The Gouy−Chapman−Stern model captures many important features of the interfacial structure, even though it does have its limitations in even moderately concentrated electrolytes [24−26]. It proposes a layered interfacial structure: an immobile inner layer referred to as the Stern layer and an outer, diffuse layer. Ions in the inner layer are considered to be bound to the surface whereas ions in the diffuse layer are mobile, described by the Boltzmann distribution with the electrical potential decreasing exponentially away from the surface. The concentration of individual ions representing the excess charge, c_i, can be described by Eq. (1.25) with c_∞ representing the bulk concentration of ions, z their valency, and Φ the electric potential:

$$c_i = c_\infty \exp\left(\frac{-ze\Phi}{k_B T} \right) \tag{1.25}$$

The Gouy−Chapman−Stern model is a useful tool for understanding the electrical double layer but fails in quantitative terms in all but the lowest electrolyte

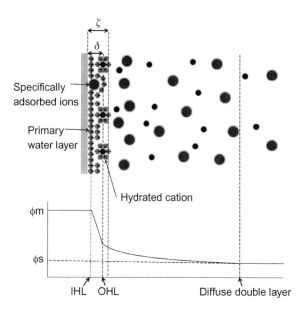

FIGURE 1.5

Ion distribution in the electric double layer at a charged solid/liquid interface with inner Stern and outer diffuse layer according to the Gouy–Chapman–Stern model, including the local potential distribution. The characteristic decay length is approximately equal to the Debye length λ_D, see below.

concentrations. This failure is rooted in the rather simplifying assumptions inherent in the model, even though it does predict trends quite accurately [27]:

- Ions in solution are considered as a continuous charge distribution with the discrete nature of the charges ignored. The surface charge of the solid is also assumed to be homogeneous when in reality it is localized with individually charged groups and adsorbed ions.
- Non-Coulombic interactions are disregarded.
- The permittivity of the solvent is taken to be constant and continuous. This is not the case at the surface where the strong electric field prevents the rotation of the polar solvent molecules and high concentrations of counterions can drastically change the permittivity.
- The surface is assumed to be flat on a molecular scale.

The key difference of the Gouy–Chapman–Stern model over previous representations of the double layer is in the consideration of the electrolyte close to the surface. In previous models based on the Poisson–Boltzmann distribution, the finite size of the ions was neglected in this area, a crude approximation as in this region the ion concentration can be very high; a surface potential of 100 mV can lead to an increase in concentration of around 50 times compared to the bulk. The Gouy–Chapman–Stern model's introduction of the Stern layer is an attempt to account for ion size effects.

In the simplest case of the Stern layer, only the finite size of the counterions themselves is considered. An ion with a finite size cannot get infinitely close to the surface and will always remain at a given distance. This distance, given the term δ, between the surface of the material and the center of the first layer of counterions marks the Helmholtz plane; the boundary between the Stern and the diffuse layers. In water, the finite size of the counterions may include their hydration shell, increasing the distance δ. The diffuse layer potential is found at the distance δ from the surface of the solid and should not be confused with the ζ-potential that is also indicated in Figure 1.5. The differences between the diffuse layer and the ζ-potentials are further discussed later in this chapter.

A more sophisticated model of the Stern layer takes the specific adsorption of ions into account [28]. Here, the Helmholtz plane can no longer be considered as a single plane. Ions that have been specifically adsorbed bind tightly to the surface, losing their hydration shells. The locus of the electrical center of these specifically adsorbed ions is defined as the inner Helmholtz plane (IHP). With a more sophisticated model of the electrical double layer, we can also define the outer Helmholtz plane (OHP), the plane of closest approach of solvated ions.

1.5.2 The conductance of nanopores with charged inner walls

In nanopores with charged inner walls, the ion distribution inside the pore is generally different from the bulk solution. As outlined above, counterions accumulate close to the walls due to the favorable electrostatic interactions with the surface, whereas co-ions are depleted. In the case of silicon oxide or silicon nitride membranes immersed in KCl electrolyte, the surface charge density σ_{surf} is typically negative due to ionized hydroxyl or oxide groups. The K^+ concentration at the interface is, therefore, increased and the local Cl^- concentration decreased. The spatial extension of the electric double layer is again characterized by the Debye length λ_D and varies from a fraction of a nanometer for concentrated solutions to tens of nanometers in very dilute electrolyte.

For larger pores or at high ionic strength, $d > \lambda_D$, the double layer structure is unperturbed compared to a single flat surface and the electrostatic potential drops to zero at a sufficiently large distance away from the pore wall. On the other hand, for very small pores where $d \sim \lambda_D$ and smaller, the electric double layer structure is affected by the small dimensions of the pore due to charge as well as curvature effects. In any case, the excess charge in the double layer will respond to an external electric field and contribute to the flux of ions in the pore. While structural details of the double layer will no doubt affect this response, for example, due to a concentration-dependent mobility or surface interactions, to a first approximation one may simply add the flux of excess charge to the bulk flux of ions.

If the surface charge density is σ_{surf}, then the total charge to be neutralized by counterions in solution is $\sigma_{surf} \cdot \pi \cdot d \cdot L$, where $\pi \cdot d \cdot L$ is the inner surface area of a cylindrical pore. The excess concentration of counterions is then $-\sigma_{surf} \cdot \pi \cdot d \cdot L$ divided by the pore volume V_{pore}, which in molar units is equal to:

$$c_{dl} = -\frac{4}{z_{dl}F} \cdot \frac{\sigma_{surf}}{d} \tag{1.26}$$

z_{dl} is the formal ionic charge of the counterion. For K^+ in KCl electrolyte in contact with a negatively charged wall, $z_{dl} = +1$. According to Eq. (1.12), the associated flux is then:

$$j_{dl} = -4D_{dl} \cdot \frac{\sigma_{surf}}{d \cdot RT} \cdot E \tag{1.27}$$

and the current density, in analogy to Eq. (1.13):

$$J = \frac{z_i^2 \cdot F^2}{RT} \cdot [D_+ + D_-] \cdot c \cdot E + z_{dl} \cdot F \cdot \left(-4D_{dl} \cdot \frac{\sigma_{surf}}{d \cdot RT} \cdot E \right) \tag{1.28}$$

For a pore with cross-sectional area $A = \pi d^2/4$, $\sigma_{surf} < 0$, $z_{dl} = 1$, and employing the Einstein relation, one obtains for the pore current I:

$$I = \frac{\pi d^2}{4L} \sigma_s \cdot V_{bias} + \pi \frac{d}{L} \cdot |\sigma_{surf}| \cdot u_i \, z_i \, e \cdot V_{bias} \tag{1.29}$$

Note that σ_s is the solution conductivity in Ω/m (not a charge density), as defined in Eq. (1.13). $|\sigma_{surf}|$ is the absolute charge density of the inner pore surface and u_i the absolute electrophoretic mobility. Accordingly, the expression for G_{pore} is shown in Eq. (1.30), which has been used in the literature to investigate surface effects [29].

$$G_{pore} = \frac{\pi d^2}{4L} \sigma_s + \pi \frac{d}{L} \cdot |\sigma_{surf}| \cdot u_i \, z_i \, e \tag{1.30}$$

G_{pore} is thus the sum of bulk and surface contributions, which represent two parallel conduction pathways. The relative surface contribution is expected to be larger for higher surface charge densities and smaller pore diameters, which is in accordance with experimental data. Figure 1.6 shows simulations based on Eq. (1.30), illustrating the effect of both parameters. As for the uncharged pore, the ion current−voltage curve is expected to be symmetric with no rectification, despite the fact that there is an imbalance between cation and anion fluxes. Importantly, this imbalance is the same, irrespective of the bias direction, and therefore the current at positive and negative biases of equal magnitude are the same. Rectification may be induced by combining geometric and electrostatic effects, for example, in narrow conical pores with surface charge, where the activation energy for ions entering the channel depends on the direction of transport [30−33].

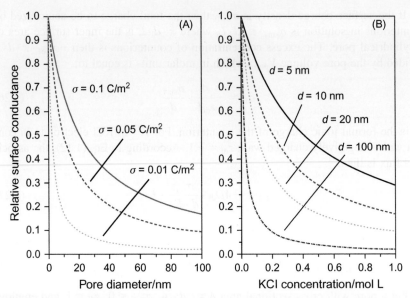

FIGURE 1.6

Surface conductance relative to G_{pore} based on Eq. (1.30): (A) as a function of diameter for different surface charge densities σ_{surf} (c(KCl) = 0.1 M); (B) as a function of KCl concentration for different pore diameters d (σ_{surf} = 0.1 C/m^2). L = 310 nm, $u_{(K+)}$ = 7.616 × 10^{-8} m^2/Vs, $u_{(Cl-)}$ = 7.909 × 10^{-8} m^2/Vs.

Source: Figure reproduced with permission from Ayub et al. [17].

1.5.3 The ζ-potential of colloids and charged particles

From the structure of the solid–liquid interface, it is clear that not all ions in the electric double layer are equally mobile. In particular, if the liquid moves relative to the solid surface, a shear plane evolves which is generally not directly at the solid surface. Only molecules at a characteristic distance ζ from the interface will start to move; the electric potential at this distance compared to the bulk fluid is known as the ζ-potential.

Based on the Gouy–Chapman–Stern model, it is often assumed that the shear plane is found at the OHP, a distance δ from the surface. Conceptually, this is not the case even though the two planes lie very close to each other [34,35]. The potential at the OHP is known as the diffuse layer potential and whilst it is very similar in value to the ζ-potential as would be expected from their close proximity, they are formally not identical to each other (Figure 1.5) [36].

The ζ-potential is a measurable quantity related to both the colloidal stability of particles in electrolyte solution and their electrophoretic mobility. The ζ-potential of an analyte is thus an important parameter when undertaking nanopore experiments. Ideally, a ζ-potential with an absolute magnitude of 30 mV or greater is required for analytes to have good stability and to prevent flocculation (irreversible aggregation) [37]. Moreover, the ζ-potential is related to the surface charge of the analyte—albeit not necessarily in a trivial way [38]—and therefore a relevant parameter in determining the direction of electrophoretic transport through the nanopore.

One of the simplest and most common approaches to determine the ζ-potential for nanopore applications involves measuring the conventional electrophoretic mobility, u_{conv}, of a particle under given solution conditions. As mentioned earlier, the conventional electrophoretic mobility relates the drift speed of a species to the applied electric field E, when turbulent flow is absent [39].

After the electrophoretic mobility of the particles has been determined, Smoluchowski's theory is used to relate u_{conv} to the ζ-potential [40].

$$u_{conv} = \frac{\varepsilon\zeta}{\eta} \text{(Helmholtz} - \text{Smoluchowski equation)} \qquad (1.31)$$

This analysis is only valid for systems with a sufficiently thin double layer, i.e., a Debye length much smaller than the particle radius R_p.

$$\frac{R_p}{\lambda_D} \gg 1 \qquad (1.32)$$

This can be considered in the context of a nanopore experiment with a 1 M KCl buffer, for which ε_r is taken to be 71 [41]. In this case, the Debye length is

$$\lambda_D = \sqrt{\frac{\varepsilon RT}{F^2 \cdot \sum c_i \cdot z_i^2}} \qquad (1.33)$$

$$\lambda_D = 2.89 \times 10^{-10} \text{ m} \approx 0.3 \text{ nm}$$

Based on this value for λ_D, the validity of the above assumption can be assessed for several biological molecules of interest as given in Table 1.1.

Here, the condition $R_p/\lambda_D \gg 1$ is satisfied in all cases, inferring Smoluchowski's approach is a valid method of calculating ζ-potentials of biological molecules under conditions that are typical of nanopore translocation experiments. Extensions of Smoluchowski's theory involving the Dukhin number are known [46] but will not be discussed in greater detail here.

Table 1.1 Stokes Radius and R_p/λ_D for Various Biomolecules of Interest

Molecule	Stokes Radius (nm)	R_p/λ_D
20 bp DNA fragment [42]	3.0	10.4
106 kbp plasmid DNA [43]	780.0	2699.0
p53 [44]	6.4	22.2
Hemoglobin [45]	3.1	10.7

1.5.4 Electroosmosis—fluid motion close to a charged wall in response to an external electric field

When a liquid is in contact with a charged surface and an electric field is applied parallel to the interface, a movement of liquid occurs adjacent to the wall, an effect known as electroosmosis [47–52]. The fixed surface charge is balanced with cations, which are in excess relative to the bulk concentration. When an external electric field is applied, cations and anions in the solution will move with the potential gradients and drag their solvation shell along. The result is a net flow of liquid in the direction of the electric field due to the excess cations.

A full theoretical description of this flow is achieved by solving the Navier–Stokes equation with suitable boundary conditions but exact analytical solutions are only possible for simple model geometries. We illustrate the approach briefly here, but further details can be found in specialized texts [53]. The Navier–Stokes equation for incompressible flow is described as:

$$\rho\left(\frac{\partial \mathbf{v}}{\partial t} + \mathbf{v} \cdot \nabla \mathbf{v}\right) = -\nabla p + \eta \nabla^2 \mathbf{u} + \mathbf{F} \tag{1.34}$$

$\mathbf{v}$ is the velocity of the fluid, ρ is fluid's density, p is the pressure, and η is the fluid's dynamic viscosity. $\mathbf{F}$ is a force term, which in the present case represents the applied external electric field, i.e., $\mathbf{F} = \rho_E \cdot E$, where ρ_E is charge density of the solution. Inside the electric double layer $\rho_E \neq 0$, since there is an excess charge present; far away from the charged surface, where the composition is equal to the bulk, $\rho_E = 0$.

Focusing on the electric double layer part first, we note that the electric double layer extends in the y-direction away from the surface. Assuming that flow occurs strictly in the x-direction along a planar charged surface in the absence of any pressure differences (isobaric flow), we can simplify Eq. (1.34) considerably: $dv/dt = 0$ (steady-state); $\nabla p = 0$ (isobaric flow); $dv/dx = 0$ and $vy = 0$ (no velocity change in the x-direction at given y; no velocity component in the y-direction, since only flow in the x-direction is allowed according to our simplified model). All derivatives in the z-direction are zero. Therefore, we have:

$$0 = \eta \nabla^2 \mathbf{v} + \rho_E \cdot E \tag{1.35}$$

The variation of the electrostatic potential across the double layer is determined by Poisson's equation:

$$-\varepsilon \nabla^2 \Phi = \rho_E \tag{1.36}$$

ε is the dielectric constant of the medium ($=\varepsilon_r \cdot \varepsilon_0$). The resulting differential equation (1.37) may be integrated:

$$\varepsilon E \frac{d^2 \Phi}{dy^2} = \eta \frac{d^2 v}{dy^2} \tag{1.37}$$

If the ζ-plane is set as $y = 0$ and the no-slip condition invoked $(\mathbf{v}_{(y=0)} = 0)$, then the fluid velocity inside the double layer is found to be (see Figure 1.5):

$$\mathbf{v}_{in} = \frac{\varepsilon E}{\eta}(\Phi - \zeta) \tag{1.38}$$

Often the assumption $\Phi_0 \approx \zeta$-potential, where Φ_0 is the surface potential, is made resulting in the equation:

$$\mathbf{v}_{in} = \frac{\varepsilon E}{\eta}(\Phi - \Phi_0) \tag{1.39}$$

Since Φ is a function of y, the velocity changes in a similar way as the potential (Figure 1.7).

In order to calculate the fluid velocity outside the double layer (far away from the wall), the Navier–Stokes equation (1.34) has to be solved with different boundary conditions. For example, since the electrolyte solution is electrically neutral, the force term $\mathbf{F} = \rho_E \cdot E$ disappears. The "outer" solution for the fluid velocity then becomes:

$$\mathbf{v}_{out} = -\frac{\varepsilon \cdot \Phi_0}{\eta} \cdot E \tag{1.40}$$

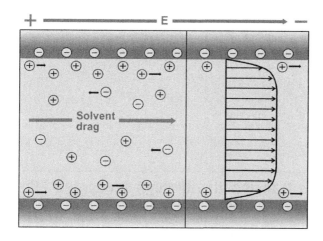

FIGURE 1.7

An illustration of electroosmotic flow close to a negatively charged solid surface. Left: Cations counterbalance the fixed surface charge and are therefore in excess relative to the bulk concentration. Upon application of an external electric field, cations move in accordance with the potential gradient and drag their solvation shells along, resulting in net fluid motion from left to right. Right: Corresponding flow profile. Under nonslip conditions, the fluid velocity at the surface is zero and increases across the double layer to reach a constant value inside the channel.

Equation (1.40) can again be used to define the electroosmotic mobility as $u_{EO} = -\varepsilon \cdot \Phi_0/\eta$, which is identical to Eq. (1.31) for the electrophoretic mobility of a colloid particle in solution if we again make the assumption $\Phi_0 \approx \zeta$-potential, the Helmholtz–Smoluchowski equation. This may appear puzzling at first but is merely a reflection of the fact that in both cases we are looking at the relative motion of a charged solid with respect to the solution or vice versa. In the case of a nanopore channel, the solid is "infinitely" large on the molecular scale and one focuses on, and potentially measures, the flow of liquid along the surface. In the case of colloids, one monitors the motion of a particle in the liquid medium. In both cases, motion is motivated by an external electric field.

Electroosmotic effects may thus play a role in nanopore sensing experiments, since electroosmosis may act in the same or the opposite direction of the electric field. Their magnitude depends on the surface properties of the pore material and the analyte, as well as the solution. Electroosmotic effects are most prominent in long nanochannels, where flow profiles have more time to fully develop [54]. This has been shown in the case of cylindrical pores in silicon nitride membranes [55].

Electroosmotic effects have not been observed in nanopore experiments with DNA, potentially due to the large fixed charge on the sample. However, proteins exhibit significantly lower effective charge, which can vary depending on solution conditions, such that electroosmosis plays a bigger role even in cases where short nanochannels are used. Indeed, for proteins a combination of electrophoretic and electroosmotic forces determine the translocation direction and time [56]. According to these authors, the effective velocity is:

$$\mathbf{v}_{\text{eff}} = \frac{\varepsilon}{\eta}(\zeta_{\text{protein}} - \zeta_{\text{pore}}) \cdot E \qquad (1.41)$$

Depending on the relative ζ-potentials of protein and pore surface, the direction of transport may thus vary.

1.6 Particle translocation through nanopores—the model of deBlois and Bean

As discussed in previous sections, the pore current is a measure for the net transport of charged species through the pore per unit time, cf. Eq. (1.8). It is therefore not surprising that other species entering the pore, such as polyelectrolytes (DNA, proteins) or colloids, affect ion transport and thus the pore current. This is the fundamental basis of electric nanopore sensing. The characteristic features of the resulting current modulation are related to the properties of the analyte as well as those of the nanopore and the solution. In a simple volume exclusion model, the analyte is simply viewed as neutral, such that insulating species replace a volume element of solution with a given ion concentration.

Large analytes, relative to the size of the pore, would thus result in a strong reduction of the pore current. The event duration would be proportional to the dimensions of the analyte in the direction of transport. On the whole, this picture is too simplified, certainly for DNA where current modulations have been shown to be both positive and negative, depending on the ionic strength of the solution [57].

Taking into account the charge of the analyte, one might expect that analytes with a small effective volume charge density—smaller than the charge density of the electrolyte that they replace—would cause current blockage, whereas highly charged analytes produce current enhancement. While this view appears to be qualitatively correct, more sophisticated models need to be applied, depending on the analyte, in order to reproduce the current modulation features with some quantitative accuracy. This is not a trivial task, however, since hydrodynamic effects (electroosmosis), molecule/surface interactions, the local charge distribution (concentration polarization, ion condensation) as well as the properties of the translocating analyte are important.

Simple models still have their merit, however, in that they illustrate the fundamental physics of the translocation process. These insights may then be applied in the context of more detailed and accurate theoretical models and computer simulations.

To this end, DeBlois and Bean [8] have developed a model that describes the resistance change that is expected for an uncharged, insulating colloidal particle that is placed within a pore. They examine this problem in two ways: first they derive an expression for small particles relative to the pore dimensions, and second they produce a wider reaching solution that is valid for a broader range of particle sizes. Both will be discussed below, while the reader is directed to DeBlois and Bean [8] for a full derivation.

1.6.1 Small spheres solution

The first consideration is of the effective resistivity ρ_{eff} of a dilute suspension of insulating spheres in solution, which was given by Maxwell [58] to be:

$$\rho_{eff} = \frac{1}{\sigma_s}\left(1 + \frac{3f}{2} + \ldots\right) \tag{1.42}$$

where f is the fraction of the total volume of the pore that is occupied by a sphere.

Consider now a cylindrical pore of diameter d and length L filled with fluid of conductivity σ_s, Figure 1.8B. For a pore which has a diameter much smaller than its length, the resistance is the inverse of Eq. (1.16):

$$R_1 = \frac{4L}{\sigma_s \pi d^2} \tag{1.43}$$

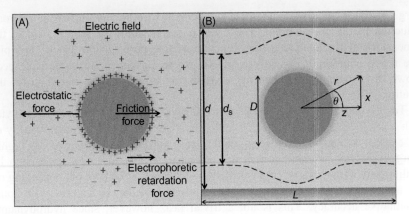

FIGURE 1.8

(A) A charged colloidal particle in solution under the influence of an external electric field (schematic). The electric field drags the particle toward the left, which is however (partly) countered by a velocity-dependent friction force and a retardation force, *vide supra*. (B) The same particle inserted in a cylindrical nanopore channel. The parameters shown are used in the derivations, as described fully by DeBlois and Bean [8]. The dotted line indicates the streamlines that develop while the particle is within the channel.

Access resistance may be taken into account such that the resulting R_{pore} can be given as:

$$R_1 = \frac{4(L + 0.8d)}{\sigma_s \pi d^2} \tag{1.44}$$

Note that an identical expression emerges from rearranging Eq. (1.24).

A sphere of diameter D is considered that will occupy a fraction of the total volume of the pore such that:

$$f = \frac{2D^3}{3d^2 L} \tag{1.45}$$

Substituting this occupied volume fraction into the effective resistivity equation above, and inserting this relation into Eq. (1.43) where $\sigma_s = 1/\rho_{\text{eff}}$, we obtain:

$$R_2 = \frac{4L}{\sigma_s \pi d^2} \left(1 + \frac{D^3}{d^2 L} + \ldots \right) \tag{1.46}$$

This second resistance describes the resistance of a pore upon insertion of a small spherical particle.

The change in resistance ΔR due to this simple volume exclusion calculation is thus:

$$\Delta R = R_2 - R_1 = \frac{4D^3}{\sigma_s \pi d^4} \tag{1.47}$$

This change in conductance only holds for spheres which have diameters that are much smaller than the pore diameter ($D \ll d$). Note the very strong dependence on both channel and particle dimensions.

1.6.2 "Broad range" solution

A complete solution that is valid over a broad range of particle sizes requires a solution of Laplace's equation for the potential, with appropriate (insulating) boundary conditions at both the sphere and the pore walls. The potential obtained from this solution corresponds to tubular current streamlines that display a slight bulge around the sphere, as shown by d_s in Figure 1.8B. The streamline varies with diameter while carrying constant current.

The change in resistance upon introduction of an insulating sphere into the pore is found to be:

$$\Delta R = R_2 - R_1 = \frac{4}{\sigma_s \pi d^2} \left[1 - \left(\frac{D}{d} \right)^3 \right]^{-1} \left[\frac{D^3}{2L^2} + 2 \int_0^{L/2} \frac{D^3 \, dz}{(d_s^2 + 4z^2)^{3/2}} \right] \qquad (1.48)$$

The upper limit of the current change, where $d_s = d$, for particle diameters much smaller than the pore diameter, $D \ll d$, is given by:

$$\Delta R_{\left(\frac{D}{d}\right) \ll 1, \left(\frac{d}{L}\right) < 1} \rightarrow \frac{4}{\sigma_s \pi d^2} \left[1 + \frac{3}{8} \left(\frac{d}{L} \right)^4 + \cdots \right] \qquad (1.49)$$

For nanopores with channel lengths greater than the channel diameter ($L > d$), Eq. (1.49) simplifies to the expression obtained by Maxwell, Eq. (1.47). Note that here ΔR measures local values of d.

For $D/d < 1$, a correction term can be applied to the Maxwellian equation such that:

$$\Delta R_{\left(\frac{d}{L}\right) < 1} = \frac{4D^3}{\sigma_s \pi d^4} \left[F\left(\frac{D^3}{d^3} \right) \right] \qquad (1.50)$$

where the correction function is given by:

$$F\left(\frac{D^3}{d^3} \right) = 1 + 1.26_8 \frac{D^3}{d^3} + 1.1_7 \frac{D^6}{d^6} \qquad (1.51)$$

Table 1.2 shows the correction function values for specific $(D/d)^3$ ratios.

The applicability of the DeBlois and Bean model can be evaluated by comparing theory with experimental results. The pore diameter (d), pore length (L), particle or analyte diameter (D), and solution conductivity were extracted from publications and applied to the corrected Maxwellian equation (Eq. (1.50)). Ratios of pore diameter to pore length and particle diameter to pore diameter are given in Table 1.3, along with the nearest corresponding value for the correction function F. The calculated value for the relative change in resistance is then

Table 1.2 Correction Function F for Different Ratios of $(D/d)^3$

$(D/d)^3$	F
0	1
0.1	1.14
0.2	1.32
0.3	1.55
0.4	1.87
0.5	2.31
0.6	2.99
0.7	4.15
0.8	6.50
0.9	13.7

Source: Table from DeBlois and Bean [8].

compared to the relative change in current from experimental data, which is expected to be similar. For the corrected Maxwellian equation (Eq. (1.50)) of the deBlois and Bean model to hold, $d/L < 1$ is required.

For the Bovine Serum Albumin (BSA), silica, and polystyrene nanoparticle experiments, this condition does not hold. Interestingly, although the DeBlois and Bean model overestimates the relative change in resistance for both the BSA and the silica nanoparticle, the estimate for polystyrene nanoparticles is a good prediction for the observed relative current change. It is expected that the DeBlois and Bean model will not hold for nanopore experiments where the pore diameter is larger than the pore length.

For experiments with Rec A and HPr proteins, where the $d/L < 1$, the estimated relative changes in resistance are of the same order of magnitude as the observed relative current changes.

In general, the $d/L < 1$ relation can be concluded to provide a limiting factor for the applicability of the DeBlois and Bean model. Where the relation holds, the DeBlois and Bean model yields the correct order of magnitude compared to the observed values. In cases where the nanopore diameter is larger than the channel length, a breakdown occurs and the corrected Maxwell equation is not applicable.

More sophisticated models are available for nonspherical particles or charged biopolymers. As DNA and proteins possess inherent substructure, they can potentially unfold upon translocation, depending on the pore dimensions and the applied field, and thus possess a more complex charge distribution [59].

Table 1.3 Calculated Relative Resistance Changes Upon Analyte Translocation Compared with Experimental Relative Current Changes for the DeBlois and Bean Model

Analyte	D (nm)	d (nm)	L (nm)	σ (S/m)	d/L	D/d	F	ΔR DeBlois and Bean	ΔI = ΔI_B/I_0
Silica nanoparticle [60]	85	175	50	0.1413	3.5	0.49	1.14	0.4572	− 0.137
Polystyrene nanoparticle [61]	22	150	50	2.24	3	0.15	1	0.0096	− 0.0081
BSA [62]	11.03	58	20	11.18	2.9	0.19	1	0.0199	− 0.0050
Rec A [63]	8.3	23.3	60	11.18	0.39	0.36	1.14	0.0200	− 0.0194
HPr [64]	2.86	7	15	11.18	0.47	0.41	1.32	0.0420	− 0.0750

BSA and RecA diameters extrapolated from crystal structure data [65,66], conductivities obtained from Sigma Aldrich standards and extrapolated for 0.2 M KCl.

BOX 1.1 LIST OF SOME PARAMETERS AND VARIABLES USED IN THIS CHAPTER

D_i	diffusion coefficient of species i, in m^2/s
$\mathbf{v}_i$	velocity of species i, in m/s; either an actual velocity or net drift velocity (v_d)
u_{abs}	absolute mobility, relating force, and speed; formally in s/kg
u_{conv}	conventional (electrophoretic) mobility, relating electric field and speed, in m^2/Vs
	In dilute solution related to the diffusion coefficient via the Einstein relation
u_{EO}	electroosmotic mobility
μ	chemical potential, in J/mol; μ^0 is the standard chemical potential
μ_{el}	electrochemical potential, in J/mol (both μ and μ_{el} are energies, not potentials!)
E	(external) electric field, in V/m
Φ	potential, in V
λ_D	Debye screening length, in m
η	viscosity of the liquid medium, in Pa s
j_i	flux of species i, in number of particles (or mol)/(m^2 s)
I	current, in A
J	current density, in A/m^2
ε	absolute permittivity of a medium (unit-less)
σ_s	solution conductivity, in $(\Omega m)^{-1}$ or S/m (S: Siemens)
G_{pore}	pore conductance, in S
ρ_s	solution resistivity, in Ω m, $\rho_s = 1/\sigma_s$
ρ_{eff}	solution resistivity in the presence of colloidal, nonconducting particles
R_{pore}	pore resistance, in Ω, $R_{pore} - (G_{pore})^{-1}$
σ_{surf}	surface charge density, in C/m^2
C	capacitance, in F (Coulomb/V)
ρ_E	charge density of the solution, in C/m^3

BOX 1.2 WHY (ALMOST) ALWAYS KCL ELECTROLYTE?

KCl is usually the electrolyte of choice in nanopore sensing experiments, particularly at high concentrations of up to 1 M and more. Such high chloride concentrations are not necessarily the best option with regard to the analyte in question, for example, DNA or proteins. The ionic strength of a solution can have a significant effect on the structure and function of biomolecules, such as changing the persistence length of DNA (by changing the screening length of electrostatic interactions) or even causing protein aggregation when van der Waals forces dominate intermolecular interactions. From a sensing point of view, it is desirable to reduce potential drops at the electrode/solution interface and inside the solution as much as possible, since then the potential drop across the nanopore is essentially equal to the externally applied bias voltage V_{bias}. This is achieved by high solution conductivity σ_s and the use of "ideal nonpolarizable" electrodes, which exhibit fast interfacial charge transfer and a high exchange current density. This is the case for Ag/AgCl electrodes in contact with a Cl^--containing electrolyte, where Ag and insoluble AgCl are interconverted. However, the same could be achieved with other reversible redox couples, such as other silver halides, phosphates, or the reversible hydrogen electrode, so why KCl in particular? KCl has another advantage in that the ion mobilities of K^+ and Cl^- in aqueous electrolyte are almost the same. If the mobilities were significantly different, one ion would move faster than the other. In the case of diffusion along a concentration gradient, both ions move in the same direction. But since their mobilities are different, a local charge imbalance and a so-called "diffusion potential" emerge (see also "liquid-junction potential"), counteracting the external diffusion field and in effect decreasing the mobility difference. Under the influence of an electric field, the two ions move in opposite directions. At the nanopore, the faster ion moves across more quickly than the less mobile counterion (from the opposite side). Again, this causes a local charge imbalance, so-called "concentration polarization," which counteracts the external electric field—a phenomenon well known in membrane separation science. Recently, it has also been suggested that concentration polarization effects could also affect the translocation characteristics of DNA, and perhaps other biopolymers [67].

References

[1] Hille B. Pharmacological modifications of the sodium channels of frog nerve. J Gen Physiol 1968;51:199−219.

[2] Hall J. Access resistance of a small circular pore. J Gen Physiol 1975;66:531−2.

[3] Hille B. Ionic channels in nerve membranes. Prog Biophys Mol Biol 1970;21:1−32.

[4] Albrecht T. How to understand and interpret current flow in nanopore/electrode devices. ACS Nano 2011;5:6714−25.

[5] Buck R. Kinetics of bulk and interfacial ionic motion: microscopic bases and limits for the Nernst−Planck equation applied to membrane systems. J Memb Sci 1984;17:1−62.

[6] Chen D, Lear J, Eisenberg B. Permeation through an open channel: Poisson−Nernst−Planck theory of a synthetic ionic channel. Biophys J 1997;72:97−116.

[7] Schoch R, Han J, Renaud P. Transport phenomena in nanofluidics. Rev Mod Phys 2008;80:839−83.

[8] deBlois R, Bean C. Counting and sizing of submicron particles by the resistive pulse technique. Rev Sci Instrum 1970;41:909.

[9] Pedone D, Langecker M, Abstreiter G, Rant U. A pore−cavity−pore device to trap and investigate single nanoparticles and DNA molecules in a femtoliter compartment: confined diffusion and narrow escape. Nano Lett 2011;11:1561−7.

[10] Cheviakov A, Reimer A, Ward M. Mathematical modeling and numerical computation of narrow escape problems. Phys Rev E 2012;85:1−16.

[11] Lu B, Albertorio F, Hoogerheide D, Golovchenko J. Origins and consequences of velocity fluctuations during DNA passage through a nanopore. Biophys J 2011;101:70−9.

[12] Turq P, Barthel J, Chemla M. Transport relaxation and kinetic processes. Berlin: Springer Verlag; 1992.

[13] Shaw D. Introduction to colloid and surface chemistry. 4th ed. Butterworth-Heinemann; 1992.

[14] Bockris J, Reddy A. In: Bockris JO, editor. Modern electrochemistry, vol. 1, Ionics. 2nd ed. New York, NY: Plenum Press; 1998 [chapter 4].

[15] Glasstone S. An introduction to electrochemistry. New York: Van Nostrand; 1942.

[16] Kowalczyk S, Grosberg A, Rabin Y, Dekker C. Modeling the conductance and DNA blockade of solid-state nanopores. Nanotechnology 2011;22:315101.

[17] Ayub M, Ivanov A, Instuli E, Cecchini M, Chansin G, McGilvery C, et al. Nanopore/electrode structures for single-molecule biosensing. Electrochim Acta 2010;55:8237−43.

[18] Siwy Z, Fuliński A. A nanodevice for rectification and pumping ions. Am J Phys 2004;72:567.

[19] Wei R, Gatterdam V, Wieneke R, Tampé R, Rant U. Stochastic sensing of proteins with receptor-modified solid-state nanopores. Nat Nanotechnol 2012;7:257−63.

[20] Smythe W. Static and dynamic electricity. McGraw-Hill, New York; 1950.

[21] Hyun C, Rollings R, Li J. Probing access resistance of solid-state nanopores with a scanning probe microscope tip. Small 2012;8:385−92.

[22] Garaj S, Hubbard W, Reina A, Kong J, Branton D, Golovchenko J. Graphene as a subnanometre trans-electrode membrane. Nature 2010;467:190−3.

[23] Schneider G, Kowalczyk S, Calado V, Pandraud G, Zandbergen H, Vandersypen L, et al. DNA translocation through graphene nanopores. Nano Lett 2010;10:3163−7.

[24] Chapman D. A contribution to the theory of electrocapillarity. Philos. Mag. 1913;25:475.

[25] Gouy G. Sur La constitution de la charge electrique a la surface D'un electrolyte. Compt Rend 1909;149:654−7.

[26] Stern O. Zur Theorie der Elektrolytischen Doppelschicht 1924;30:508.

[27] Cevc G. Membrane electrostatics. Biochim Biophys Acta 1990;1031:311−82.

[28] Yates D, Levine S, Healy T. Site-binding model of the electrical double layer at the oxide/water interface. J Chem Soc Faraday Trans 1 1974;70:1807−18.

[29] Hall A, van Dorp S, Lemay S, Dekker C. Electrophoretic force on a protein-coated DNA molecule in a solid-state nanopore. Nano Lett 2009;9:4441−5.

[30] He Y, Gillespie D, Boda D, Vlassiouk I, Eisenberg R, Siwy Z. Tuning transport properties of nanofluidic devices with local charge inversion. J Am Chem Soc 2009;131:5194−202.

[31] Howorka S, Siwy Z. Nanopore analytics: sensing of single molecules. Chem Soc Rev 2009;38:2360−84.

[32] Baris B, Jeannoutot J, Luzet V, Palmino F, Rochefort A, Cherioux F. Non-covalent bicomponents self-assemblies on a silicon surface. ACS Nano 2012;28(6):6905−11.

[33] Constantin D, Siwy Z. Poisson−Nernst−Planck model of ion current rectification through a nanofluidic diode. Phys Rev E 2007;76:1−10.

[34] Chan D, Horn R. The drainage of thin liquid films between solid surfaces. J Chem Phys 1985;83:5311.

[35] Israelachvili J. Measurement of the viscosity of liquids in very thin films. J Colloid Interface Sci 1986;110:263−71.

[36] Attard P, Antelmi D. Comparison of the zeta potential with the diffuse layer potential from charge titration. Langmuir 2000;1542−52.

[37] Sherman P. Rheology of disperse systems. Industrial rheology. London: Academic Press Inc. P. Sherman, Trade Cloth; 1970, 97−183.

[38] Salis A, Boström M, Medda L, Cugia F, Barse B, Parsons D, et al. Measurements and theoretical interpretation of points of zero charge/potential of BSA protein. Langmuir 2011;27:11597−604.

[39] Lyklema J. On the slip process in electrokinetics. Colloids Surf A 1994;92:41−9.

[40] Smoluchowski M. Contribution a la theorie de L'endosmose electrique et de quelques phenomenes correlatifs. Bull Int Acad Sci Cracov 1903;8:182−200.

[41] Wu J, Stark J. Measurement of low frequency relative permittivity of room temperature molten salts by triangular waveform voltage. Meas Sci Technol 2006;17:781−8.

[42] Wanunu M. Nanopores: a journey towards DNA sequencing. Phys Life Rev 2012;9:125−58.

[43] Araki S, Nakai T, Hizume K, Takeyasu K, Yoshikawa K. Hydrodynamic radius of circular DNA is larger than that of linear DNA. Chem Phys Lett 2006;418: 255−9.

[44] Friedman P, Chen X, Bargonetti J, Prives C. The P53 protein is an unusually shaped tetramer that binds directly to DNA. Proc Natl Acad Sci USA 1993;90:3319−23.

[45] Doster W, Longeville S. Microscopic diffusion and hydrodynamic interactions of hemoglobin in red blood cells. Biophys J 2007;93:1360−8.

[46] Dukhin S. Non-equilibrium electric surface phenomena. Adv Colloid Interface Sci 1993;44:1−134.

[47] Schmid G, Schwarz H. Electrochemistry of capillary systems with narrow pores. I. Overview. J Memb Sci 1998;150:151−7.

[48] Schmid G, Schwarz H. Electrochemistry of capillary systems with narrow pores. II. Electroosmosis. J Memb Sci 1988;150:159−70.

[49] Schmid G, Schwarz H. Electrochemistry of capillary systems with narrow pores. III. Electrical conductivity. J Memb Sci 1998;150:171−87.

[50] Schmid G, Schwarz H. Electrochemistry of capillary systems with narrow pores. IV. Dialysis potentials. J Memb Sci 1998;150:189−96.

[51] Schmid G, Schwarz H. Electrochemistry of capillary systems with narrow pores V. Streaming potential: Donnan hindrance of electrolyte transport. J Memb Sci 1998;150:197−209.

[52] Schmid G, Schwarz H. Electrochemistry of capillary systems with narrow pores. VI. Convection conductivity (theoretical considerations). J Memb Sci 1998;150:211−25.

[53] Kirby J. Micro and nanoscale fluid mechanics. Cambridge: Cambridge University Press; 2010 [chapter 6].

[54] Yang R, Fu L, Hwang C. Electroosmotic entry flow in a microchannel. J Colloid Interface Sci 2001;244:173−9.

[55] Yusko E, An R. Electroosmotic flow can generate ion current rectification in nano- and micropores. ACS Nano 2010;4:477−87.

[56] Firnkes M, Pedone D, Knezevic J, Döblinger M, Rant U. Electrically facilitated translocations of proteins through silicon nitride nanopores: conjoint and competitive action of diffusion, electrophoresis, and electroosmosis. Nano Lett 2010;10:2162−7.

[57] Smeets R, Keyser U, Krapf D, Wu M-Y, Dekker N, Dekker C. Salt dependence of ion transport and DNA translocation through solid-state nanopores. Nano Lett 2006;6:89−95.

[58] Maxwell J. A treatise on electricity and magnetism vol. 1. 3rd ed. Oxford: Clarendon; 1904.

[59] Ai Y, Qian S. Electrokinetic particle translocation through a nanopore. Phys Chem Chem Phys 2011;13:4060−71.

[60] Bacri L, Oukhaled A, Schiedt B, Patriarche G, Bourhis E, Gierak J, et al. Dynamics of colloids in single solid-state nanopores. J Phys Chem B 2011;115:2890−8.

[61] Prabhu AS, Jubery TZ, Freedman K, Mulero R, Dutta P, Kim MJ. Chemically modified solid state nanopores for high throughput nanoparticle separation. J Phys Condens Matter 2010;22:454107.

[62] Han A, Schürmann G, Mondin G, Bitterli RA, Hegelbach N, de Rooij N, et al. Sensing protein molecules using nanofabricated pores. Appl Phys Lett 2006;88: 093901.

[63] Smeets R, Kowalczyk S, Hall A, Dekker N, Dekker C. Translocation of reca-coated double-stranded DNA through solid-state nanopores. Nano Lett 2009;9:3089−96.

[64] Stefureac R, Trivedi D, Marziali A, Lee J. Evidence that small proteins translocate through silicon nitride pores in a folded conformation. J Phys Condens Matter 2010;22:454133.

[65] Bujacz A. Structures of bovine, equine and leporine serum albumin. Acta Crystallogr D Biol Crystallogr 2012;68:1278−89.

[66] Bell CE. Structure and mechanism of Escherichia coli RecA ATPase. Mol Microbiol 2005;58:358−66.

[67] Das S, Dubsky P, Berg A, van den Eijkel J. Concentration polarization in translocation of DNA through nanopores and nanochannels. Phys Rev Lett 2012;108:1−5.

DNA Translocation

Oliver Otto[1] and Ulrich F. Keyser[2]

[1]*Biotechnology Center, Technical University Dresden, Tatzberg Dresden, Germany;*
[2]*Cavendish Laboratory, University of Cambridge, Cambridge, UK*

CHAPTER OUTLINE

2.1 Introduction

Following up on the previous chapter discussing the relevant electrokinetic phenomena, we will now focus on the interplay between the nanopore and the translocating object. This can be a charged biological macromolecule like DNA or a

Engineered Nanopores for Bioanalytical Applications.

protein, as well as uncharged macromolecules like polyethylene glycol. We will describe the physics of the translocation process of single DNA molecules. There are several reasons for this, one is that DNA is by far the best studied molecule and long double-stranded DNAs extracted from λ-phages are the gold standard for characterizing novel nanopores for biosensing [1−7]. Extended DNA can be considered as a polyelectrolyte, representing a long molecule with fixed charge. In addition, DNA translocation is of major importance in a manifold of future applications, most prominently for nanopore-based DNA or RNA sequencing, detection of epigenetic modifications, and structural protein analysis.

In the previous chapter, it was already introduced that the electrohydrodynamic properties of any charged surface in contact with an electrolyte is determined by the spatial distribution of ions around its surface [8]. The importance of this electrical double layer (EDL) cannot be overemphasized for the understanding of all electrokinetic phenomena. In fact, as we will show and discuss here, electrophoretic migration is the key driving force for the translocation of DNA across nanopores. This was discussed in many recent publications in the literature and leads to interesting phenomena [9,10].

We will now start our description with an introduction of the relevant numbers. For the discussion in the present chapter, we assume a negatively charged surface which is a valid assumption for double-stranded DNA molecules (−0.96 n/cm) and most solid-state nanopores in solutions of monovalent ions. The solid−liquid interface leads to an enhancement in the cation concentration (positive ions with charge $+Ze^-$) and depletion of coions (negative ions with charge $-Ze^-$) in the vicinity of the negative surface. Here, e^- represents the elementary charge of an electron and Z the valency of the ions. Please note that the bulk solution will remain neutral. It is interesting to note that this molecular capacitor model was already developed by Helmholtz in 1879. However, the distribution of charges is often more complex requiring the incorporation of statistical physics. In the liquid phase, electrostatic forces compete with diffusive forces leading to a smeared-out layer. This model was developed by Gouy and Chapman in 1910 [11] and in 1942 Stern introduced a unified theory combining it with the molecular capacitor approach. He took into account that close to the charged surface the cations are immobilized due to the strong electrostatic interactions. This Stern layer has a typical thickness of a few ionic diameters. In some distance, the influence of Brownian motion increases and the ion distribution follows the Boltzmann law forming a diffusive layer.

We start with an introduction of the physics of a polyelectrolyte inside a nanopore. We introduce relevant forces as well as electrostatic phenomena governing the interaction between the translocating object and the nanopore. While this is done for model geometries of cylindrical shape, other geometries like hyperbolic nanopores were recently discussed in the literature [12]. Surprisingly, the cylindrical shape is even a good model for funnel-shaped nanopores, e.g., made of glass capillaries that we discuss at the end of this chapter. As will become clear in the next paragraphs, the high surface charge of the double-stranded DNA molecule allows the use of the linearized Poisson−Boltzmann model only in certain situations and thus we will conclude this chapter by comparing our models with finite element calculations. The latter are the only possibility to make an attempt toward quantitative comparison of experiment and theory which is not relying on all atom molecular dynamics calculations.

2.2 Physics of a polyelectrolyte inside a nanopore
2.2.1 Electrostatic potential around a charged surface

An elementary understanding of polyelectrolytes in confinement is best developed based on simple geometric models. Figure 2.1A shows the sketch of a cylindrical nanopore with negative surface charge σ_n which is filled with an electrolyte of concentration n_0 and permittivity $\varepsilon = \varepsilon_r \varepsilon_0$ (with $\varepsilon_r = 80$ being the relativity permittivity of the electrolyte and $\varepsilon_0 = 8.854 \times 10^{-12}$ As/V/m the permittivity of free space). In our following discussion, the nanopore has a radius r_n.

Here, we assume that DNA as a polyelectrolyte can be modeled as a rod of radius r_0 and surface charge σ_0 as indicated in Figure 2.1B. Like for the nanopore, we assume that the rod is immersed in an electrolyte of concentration n_0 and having a permittivity ε as introduced above. Thus, both objects can be treated in the same theoretical approach. For the nanopore and the polyelectrolyte, the net electric charge can be related to the local average electrostatic potential by the Poisson equation:

$$\nabla^2 \Phi(r) = -\frac{\rho(r)}{\varepsilon} \tag{2.1}$$

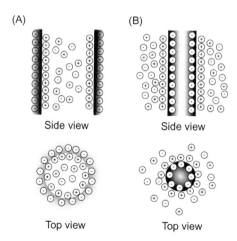

FIGURE 2.1

Ionic distribution of the positive counterions around a cylindrical surface of (A) a negatively charged nanopore and (B) a negatively charged polyelectrolyte. The physics of both objects can be analyzed by the very same fundamental laws and differs only by the applied boundary conditions.

where $\rho(r) = e[n_+ - n_-]$ is the charge density inside the electrolyte and:

$$n_\pm = n_0 e^{\mp(e\Phi(r)/k_B T)} \tag{2.2}$$

is the number density of the counterions n_+ and coions n_-, while k_B is the Boltzmann constant and T is the absolute temperature. Please note that in bulk solution $n_+ = n_- = n_0$ for a symmetric electrolyte in order to ensure electroneutrality.

Utilizing the governing relations of the Gouy−Chapman theory, Eqs (2.1) and (2.2) leads to the Poisson−Boltzmann equation (PBE):

$$\nabla^2\Phi(r) = -\frac{en_0}{\varepsilon}\left[e^{-(e\phi(r)/k_B T)} - e^{(e\phi(r)/k_B T)}\right]$$

$$\nabla^2\Phi(r) = \frac{2en_0}{\varepsilon}\sinh\left[\frac{e\phi(r)}{k_B T}\right] \tag{2.3}$$

Equation (2.3) describes the electrostatic potential for any geometry under the following assumptions:

- ions are point-like charges
- counter- and coions carry the same charge (symmetric electrolyte)
- ions interact only by Coulomb forces
- permittivity is constant inside the entire double layer
- the solvent is uniform
- the whole system is charge neutral.

The PBE is a nonlinear differential equation which can in general only be solved numerically. This is a direct consequence of the sinh-function on the right-hand side in Eq. (2.3). This complicates both analytical solutions, for many geometries a Green's function is simply not available, and the interpretation of the results is difficult. However, it is also important to notice that despite this complication, the physical principles can very often be understood by investigating the solution of the linearized equation. It might be interesting to note here as well that a number of extensions to the PBE exist that take into account a finite size of ions and other complications. A very good introduction of the extended solutions can be found in the literature [13]. In practice, many of these complications regarding higher-order corrections are sometimes taken into account by adjusting the zeta-potential of the DNA during the translocation process. We will discuss this below in more detail. It is important to note that entropy is included in this approach through the Boltzmann factor for the ion distribution. This means of course that not only Coulomb forces but also entropic contributions are taken into account. The interplay of these two effects will be investigated in Section 2.4.

The PBE has only analytical solutions for certain geometries. In order to help the basic understanding, we will now start with a major simplification. The Debye−Hueckel approach linearizes the PBE assuming a small average electrostatic potential compared to the thermal energy for situations when $e\phi/k_B T \ll 1$. In this approximation, a Taylor expansion of Eq. (2.2) leads to:

$$n_\pm = n_0\left(1 \mp \frac{e\Phi(r)}{k_B T}\right) \tag{2.4}$$

Combining Eqs (2.1) and (2.4) results in the linearized PBE:

$$\nabla^2 \Phi(r) = \kappa^2 \Phi(r) \tag{2.5}$$

which can be solved analytically, e.g., in cylinder geometry. Here the Debye length $1/\kappa$:

$$\kappa^2 = \frac{2e^2 n_0}{\varepsilon k_B T} \tag{2.6}$$

is introduced which—as a fundamental parameter—is a function of the electrolyte concentration n_0 only. In the mean-field, formalism of the Poisson–Boltzmann theory $1/\kappa$ represents a screening length where the electrostatic interactions between the surface charge and the adjacent ions are reduced to a factor $1/e$. We will see later that the Debye–Hueckel length defines one of the important length scales in the system. For an intuitive understanding, it is essential to know $1/\kappa$ in certain situations and electrolyte conditions. In pure water at pH $= 7$, $1/\kappa$ is calculated to ≈ 1 μm whereas in a 0.1 M monovalent salt we find $1/\kappa \approx 1$ nm.

We would like to emphasize that both geometries described in Figure 2.1A and B possess the same radial symmetry and can be treated simultaneously. We will assume that the system is translationally invariant. Thus, in this model the electrostatic potential $\phi(r)$ depends only on the radial coordinate r and Eq. (2.5) reduces to:

$$\frac{1}{r}\frac{\partial}{\partial r}\left(r\frac{\partial \Phi(r)}{\partial r}\right) = \kappa^2 \phi(r) \tag{2.7}$$

The actual shape of $\phi(r)$ is determined by the appropriate boundary conditions for the nanopore and DNA, respectively. Equation (2.7) has the following general solution:

$$\Phi(r) = AI_0(\kappa r) + BK_0(\kappa r) \tag{2.8}$$

where I_0 and K_0 are first-order modified Bessel functions of first and second kind, respectively. The constants A and B are determined from the boundary conditions. In order to increase the accessibility of the discussion, the solution of the nanopore and polyelectrolyte will be treated separately.

2.2.1.1 Nanopore

In case of the nanopore, as shown in Figure 2.1A the inner surface is in contact with the electrolyte. The boundary conditions inside the cylinder are:

$$\frac{d\phi(r)}{dr}\Big|_{r=0} = 0 \tag{2.9}$$

and

$$\frac{d\phi(r)}{dr}\Bigg|_{r=r_n} = \frac{\sigma_n}{\varepsilon} \tag{2.10}$$

where $r = 0$ and $r = r_n$ represent the center and surface of the nanopore, respectively. The first boundary condition reveals that $B = 0$ while constant A can be calculated from the second one. Thus the electrostatic potential inside a nanopore is represented by:

$$\phi(r) = \frac{\sigma_n}{\varepsilon\kappa}\frac{I_0(\kappa r)}{I_1(\kappa r_n)} \tag{2.11}$$

where I_1 is a second-order modified Bessel function of first kind.

2.2.1.2 Polyelectrolyte

Figure 2.1B shows the polyelectrolyte in contact with the electrolyte solution. Here, we assume the following boundary conditions:

$$\frac{d\phi(r)}{dr}\Big|_{r\to\infty} = 0 \tag{2.12}$$

and

$$\frac{d\Phi(r)}{dr}\Bigg|_{r=r_0} = -\frac{\sigma_0}{\varepsilon} \tag{2.13}$$

Using the first boundary condition in Eq. (2.8) gives $A = 0$ and from the second boundary condition it follows that:

$$\phi(r) = \frac{\sigma_0}{\varepsilon\kappa}\frac{K_0(\kappa r)}{K_1(\kappa r_0)} \tag{2.14}$$

where K_1 is a second-order modified Bessel function of second kind.

We find in both geometries that the electrostatic potential is a function of the surface charge and the permittivity as well as the salt concentration of the solution, as expected. The functional dependence is given by the appropriate modified Bessel functions. Figure 2.2A shows the potential inside a nanopore of $r_n = 100$ nm for a Debye length $1/\kappa$ of 1 nm (solid line) and 10 nm (dotted line). We find that with increasing distance from the surface the electrostatic potential decreases to $\phi(r) = 0$ in the center of the nanopore. From Eq. (2.11) it is obvious that the absolute value of the potential increases with the surface charge. Figure 2.2B shows the electrostatic potential for a polyelectrolyte with radius $r_0 = 1$ nm immersed in an ionic solution of $1/\kappa = 1$ nm (solid line) and $1/\kappa = 10$ nm (dotted line). The calculation was carried out based on Eq. (2.14) showing that $\phi(r)$ decreases with increasing distance from the charged surface.

The examples shown here are calculated for the case that the Debye screening length $1/\kappa \ll r_n$. For very small nanopores or low electrolyte concentrations, the double layer of the nanopore walls overlap and the condition of bulk neutrality is not valid any more. This leads to corrections of the nanopores conductance as shown

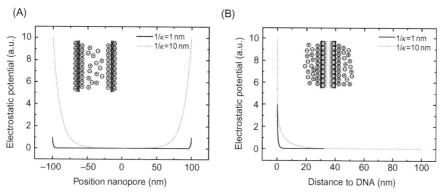

FIGURE 2.2

Electrostatic potential for two different values of $1/\kappa = 1$ nm (solid line) and $1/\kappa = 10$ nm (dotted line) calculated (A) inside a nanopore of radius $r_n = 100$ nm and (B) outside a polyelectrolyte of radius $r_0 = 1$ nm. In both cases, a constant surface charge of σ_n and σ_0, respectively was assumed. The insets show the respective geometries as introduced in Figure 2.1.

experimentally for both nanochannels [14] and nanopores [4]. Below the results for nanopores are discussed in more detail (see Figure 2.5). It is also interesting to note that when the Debye layers overlap and the surface charge of the nanopore exceeds the DNA charge, DNA cannot translocate and hence sensing is not possible. This illustrates the need for a control of the surface charge and chemistry of nano- pores. The most interesting recent approach for this is the use of lipid bilayers [15], which can be extended to other material systems like glass nanopores [16].

2.3 Electroosmotic flow inside a cylindrically nanopore

One complication of the nanopore–DNA system stems from the fact that all sur- faces are charged giving rise to a variety of electrokinetically driven flows caused by the presence of a number of charged surfaces, namely the translocating molecule and the nanopore. In our system, we apply an external potential $\Psi(x)$ across the nanopore, which leads to an electrostatic force acting on the ions (Figure 2.3). In absence of a charged surface, the motion of positive ions would be compensated by the motion of negative ions. Thus, in bulk solution there is no effective mass transport. Interestingly, when the DNA molecule is stalled or slowed down in the nanopore, electroosmotic and electrophoretic effects are not necessarily distinguish- able. This is another major factor which makes the system so interesting to study.

Inside a nanopore, the situation differs from bulk behavior. In the EDL of the surface we find an excess of counterions. Assuming a negative surface charge these positive ions (Figure 2.1B) will experience an electric body force toward the negative electrode and Stokes drag pulls adjacent water molecules with them. This leads to a plug-like flow profile and fluid is effectively pumped across the

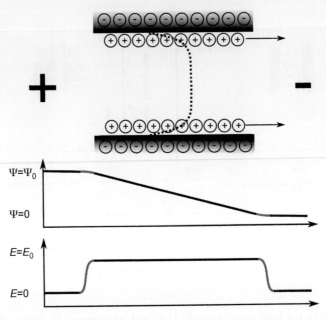

FIGURE 2.3

Fluid flow inside a cylindrical nanopore after applying an external potential Ψ_0. The positive counterions next to the charged surface drag adjacent water molecules toward the negative electrode. This gives rise to a fluid flow with a plug-like flow profile, as shown on the top. The corresponding potential and electric field E are shown below.

pore which was in detail discussed in the previous chapter. In this discussion, we assume that all shear is generated in the Debye screening layer. It is worth to notice that this is a major simplification for nanopore as the flow profile is significantly altered beyond the Debye length [9].

In general, flow profiles are described by the Navier–Stokes equation. However, in our system we can assume that the fluidic system is incompressible and inertia is safely neglected. This is allowing us to simplify the description and only the Stokes part of the equation remains:

$$\nabla^2 v + \frac{\rho(r)}{\eta} E(x) = 0 \tag{2.15}$$

with v being the fluid velocity, η its viscosity, and $E(x) = -\nabla\psi$ the external electric field in axial direction of the cylinder. Inserting Eqs (2.4) and (2.11) into Eq. (2.15) leads to the following relation describing the electrohydrodynamic problem in cylindrical coordinates:

$$\frac{1}{r}\frac{\partial}{\partial r}\left(r\frac{\partial v_x(r)}{\partial r}\right) - \frac{\kappa\sigma}{\eta}\frac{I_0(\kappa r)}{I_1(\kappa r_n)}E(x) = 0 \tag{2.16}$$

Here we used the fact that the velocity changes only in radial direction of the cylinder. Solving this equation with the appropriate boundary conditions yields:

$$v_x(r) = -\frac{1}{4}\frac{\kappa\sigma}{\eta}\frac{(r_n^2 I_0(\kappa r_n) - r^2 I_0(\kappa r))}{I_1(\kappa r_n)} E(x) \tag{2.17}$$

Figure 2.4 shows the velocity distribution inside a cylindrical nanopore. While the radius r_n was kept constant, Eq. (2.17) was evaluated for three different salt concentrations $n_{0,1}$, $n_{0,2}$, and $n_{0,3}$ representing three different Debye lengths $1/\kappa_1$, $1/\kappa_2$, and $1/\kappa_3$. For a given concentration, the velocity exhibits a plug-like profile almost across the entire nanopore. A variation in magnitude is only observed within the EDL close to the pore walls. As a consequence, the shear force on a molecule should not show any dependency on its position inside the nanopore as observed in Poiseuille flows. It is important to note here that upon inspection of Figure 2.4 the flow profile in the center of the nanopore is not completely flat. This is an important factor in determining the forces on DNA during translocation as one often finds the notion that the plug profile has a constant velocity across the nanopore (see Figure 2.2). It is obvious from Figure 2.4 that the velocity across the nanopore is not constant but varies considerably especially for small nanopores or large Debye length. This demonstrates that hydrodynamic effects are long ranged, especially when compared to the Debye length which is for the three cases of same size or 10 and 100 times smaller than the nanopore radius.

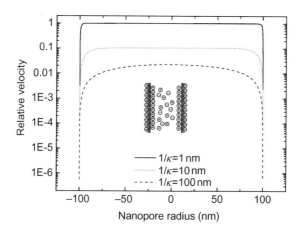

FIGURE 2.4

Velocity profile inside a nanopore of radius $R = 100$ nm. The velocity was calculated as a function of radius for three different Debye lengths of $1/\kappa = 1$ nm (solid line), $1/\kappa = 10$ nm (dotted line), and $1/\kappa = 100$ nm (dashed line). The graph shows that the magnitude in velocity only changes across a small region close to the nanopore walls for small $1/\kappa$. However, for $1/\kappa = 100$ nm the flow velocity is much lower than expected from a simple plug flow. The inset shows the nanopore geometry as introduced in Figure 2.1A.

2.4 DNA inside a nanopore

2.4.1 Free translocation

The previous chapter introduced how a nanopore can be characterized by its conductance G_{pore}. Now we will discuss the change of the nanopore conductance that indicates the presence of a DNA molecule. We will restrict the discussion to the simplest case of describing the DNA and nanopore as concentric cylinders as this covers the fundamental aspects and illustrates the physical principles. We will also have to discuss the nanopore conductance as a function of salt concentration as the variation of the Debye length is one of the easiest parameters that can be varied.

For a cylinder-like geometry, Smeets et al. [4] showed that the total current consists of a contribution of the bulk concentration of ions and the counterions shielding the surface:

$$G_{pore} = \frac{\pi r_n^2}{L}\left((\mu_+ + \mu_-)n_0 e + \mu_+ \frac{2\sigma}{r_n}\right) \tag{2.18}$$

where L is the total length of the nanopore, σ_s is the specific conductivity of the solution, and μ_+ and μ_- are the electrophoretic mobility of the counterions and coions, respectively. This is a greatly simplified yet successful approach to describe the conductance of nanopores.

Equation (2.18) defines the conductance as a function of the salt concentration n_0. In the remaining part of this chapter, we will use potassium chloride (KCl) solution as the electrolyte having electrophoretic mobilities of $\mu_+ = \mu_K = 7.6 \times 10^{-8}\,m^2/Vs$ and $\mu_- = \mu_{Cl} = 7.9 \times 10^{-8}\,m^2/Vs$. The first term in Eq. (2.18) resembles the bulk conductance of KCl while the second one represents the contribution of the surface charge of the nanopore. At sufficiently high salt concentrations, $n_{KCl} \gg \sigma/r_n e$, the first term dominates the conductance and the bulk value is obtained. With decreasing n_{KCl}, deviations from bulk behavior are observed as surface charge effects start to govern the nanopore conductance [4].

Figure 2.5A shows the conductance of 10 nm pores as a function of KCl concentration. In agreement with calculations from Eq. (2.18), the linear bulk behavior is observed for concentrations exceeding 120 mM KCl. In the low salt regime, the theory predicts a constant pore conductance which is in contradiction with experimental data (dashed line in Figure 2.5A). This disagreement is originated from the fact that in contrast to Eq. (2.18) the surface charge is a function of the ion concentration as well.

Taking this into account, Smeets et al. [4] developed a model which describes very well the nanopore conductance for KCl concentrations between 1 μM and 1 M. This is represented by the dotted line in Figure 2.5A.

The appearance of a polyelectrolyte inside the nanopore impacts on the physical description of the problem. The transport of any molecule through confinement

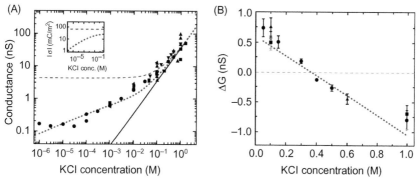

FIGURE 2.5

(A) Nanopore conductance as function of KCl concentrations from 1 μM to 1 M. (B) Conductance change due to DNA translocation for KCl concentrations from 50 mM and 1 M.

Source: Reprinted with permission from Ref. [4]. Copyright 2006. American Chemical Society.

is always driven by the complex interplay of electrostatics and hydrodynamics. Surface charge on channel walls is the reason for electroosmosis, an effect which was discussed previously. It was also shown that the effective surface charge depends on the number of counterions in the vicinity of the nanopore surface. Hence, electroosmosis can be tuned via the ionic strength of the buffer solution and its pH [9,17].

Upon translocation of a polymer, a part of the nanopore cross section will be blocked. Intuitively, the finite volume of the polyelectrolyte results in a decrease in conductance as less charge carriers are available for transport. This effect dominates at salt concentration well above 400 mM (Figure 2.5B). In experiments one finds an increase in conductance ΔG at low molarities. This can be understood by considering the counterions of the DNA backbone which introduce additional charge carriers into the nanopore [18–20] resulting in an increased ionic current.

Taking into account both effects, Smeets et al. [4] derived an expression for ΔG for translocating DNA as a function of salt concentration:

$$\Delta G = \frac{1}{L}(-\pi r_0^2(\mu_K + \mu_{Cl})n_{KCl}\, e + \mu_K^* q_{l,\,DNA}^*) \tag{2.19}$$

where r_0 is the DNA radius, μ_K^* is the effective electrophoretic mobility of potassium ions on the DNA surface, and $q_{l,DNA}^*$ is the effective DNA surface charge per unit length. Again, the first term in Eq. (2.19) represents the bulk behavior for sufficient high salt concentrations where we find $\Delta G < 0$ for translocating DNA (Figure 2.5B). When the second term becomes bigger than the first one, DNA adds counterions to the current and thus leads to a positive conductance change $\Delta G > 0$. It is important to note that Eq. (2.19) is only valid for nanopores where length and diameter are of the same order of magnitude. More general models that can be employed for a range of nanopore dimensions in both size and length were recently developed in the literature [12]. However, for the sake of simplicity and clarity we will ignore the more general and numerical treatments and the interested reader is referred to the published work in the literature (see e.g., [21−23]).

The measurements in Figure 2.5B show a linear dependency between conductance change and salt concentration. A linear fit allows to calculate the effective charge of DNA per unit length of $q^*_{l,\text{DNA}} = 0.5 e^-/$base pair[4]. However, this effective charge is depending on the diameter of the nanopore and on the salt concentration. It is important to note here that the effective charge is a nonuniversal parameter. This will be explained in more detail when discussed in the context of stalled DNA translocation using optical tweezers. These experiments clearly show that the stalling force and thus the effective charge depend on the nanopore geometry [10]. The ambiguity of the term effective charge is discussed in detail in the tutorial review by Keyser et al. [9].

2.5 Capture rate and probability of successful translocation

2.5.1 Dominating effects

Until now we were mainly concerned with understanding the distribution of ions, the resulting electroosmotic flows and the force of the molecule stalled in a nanopore. Following our focused discussion on the electrostatic and mechanical interaction between polymer and nanopore we will now analyze the crucial aspect of how DNA can enter the nanopore. We will review a few aspects of the capture process and develop a simple description of the physics. A very good treatment and description of this problem can be also found in the book by Muthukumar [24].

From a physical point of view, the capture process can be characterized by a capture rate J_c which is the fraction of polymers approaching and successfully translocating the pore. The capture rate depends on a manifold number of parameters like polyelectrolyte length and concentration, nanopore geometry and surface chemistry, salt concentration, and pH as well as the applied potential Ψ_0. Figure 2.6 displays the four steps of polymer translocation across a nanopore. In the bulk of the reservoirs, the polyelectrolyte can be assumed to diffuse freely in the absence of any pressure gradient (time t_0 in Figure 2.6). Again we can assume that far from the nanopore, the potential is constant and thus the electric field is zero. This can be understood when considering that the nanopore is the biggest resistor in the system. The electric field E will only extend to a distance of a few micrometers outside the constriction zone. Only if the polymer approaches the nanopore entrance to a distance smaller than the capture radius r_c will the motion be governed by drift (time t_1 in Figure 2.6). This drift is mainly due to the interplay of electrophoretic and electroosmotic forces and can be either repelling or attractive toward the entrance of the nanopore.

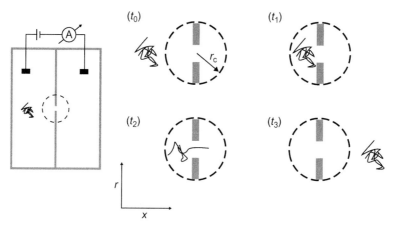

FIGURE 2.6

Complete translocation process of a polyelectrolyte through a nanopore. At t_0 the polymer diffuses freely in one reservoir. Once inside the capture radius r_c the polymer experiences a force due to electrophoretic and electroosmotic effects and drifts toward the pore (t_1). The passage requires a conformational change (t_2). After translocation the polymer is released into the reservoir (t_3).

Source: Adapted from Ref. [25].

Entering the nanopore requires a conformational change of the polyelectrolyte from a coil in thermal equilibrium with maximum entropy to an elongated state where one end of the molecule is freely accessible. This is shown in time step t_2 in Figure 2.6. The uncoiling of the polyelectrolyte chain requires overcoming a free energy barrier. Once inside, the polyelectrolyte translocates through the nanopore and is released in the reservoir (t_3 in Figure 2.6).

Translocation experiments using double-stranded DNA showed that the capture rate J_c increases first exponentially with Ψ_0 which is followed by a linear regime for higher values of the applied potential [26].

For a physical description of the problem, let us consider the number concentrations $n(x,t)$ of polymers in one dimension:

$$\frac{\partial n(x,t)}{\partial t} = -\frac{\partial}{\partial x} J(x,t) \tag{2.20}$$

where $J(x,t)$ is the net flux of molecules at coordinate x and time t. This flux consists of convective and diffusive parts:

$$J(x,t) = -D\frac{\partial n(x,t)}{\partial x} + n(x,t)v(x,t) \tag{2.21}$$

In Eq. (2.20), D is the diffusion coefficient of the polymer and $v(x,t)$ is its velocity in x direction at time t. In the following text, we will focus on the one-dimensional flux which is justified as a full radial description would change the complexity but not the physics of the problem.

In the absence of any pressure gradient, the velocity $v(x,t)$ of a protein approaching the nanopore has three components. The first contribution originates from the external potential leading to a electrophoretic velocity $v_{ep} = \mu E$. Second, gradients in the free energy landscape result in a force on the polymer and thus to a drift which scales with the hydrodynamic friction coefficient γ. Finally, electroosmosis of the electrolyte due to the nanopore surface charge directly impacts on the motion of the polymer. This can be summarized in the following equation:

$$v(x) = v_{eo} + \mu E(x) - \frac{1}{\gamma} \frac{\partial f(x)}{\partial x} \tag{2.22}$$

where v_{eo} is the electroosmotic velocity and $f(x)$ is the free energy. In the following text, we will discuss the problem in absence of electroosmosis. However, the effect of surface charge can be added by applying the derivations given in Section 2.2.

Combining Eqs (2.21) and (2.22) results in a stationary description of the flux $J(x)$:

$$J(x) = -D \frac{\partial n(x)}{\partial x} - n(x)\mu \frac{\partial \psi(x)}{\partial x} - n(x) \frac{D}{k_B T} \frac{\partial f(x)}{\partial x} \tag{2.23}$$

Here, we took into account the Einstein relation for the diffusion coefficient and that the potential $\psi(x)$ drops linear in x. The structure of Eq. (2.23) reveals the components of the flux which originate from diffusion, drift, and free energy. In the following sections these three contributions will be discussed.

2.5.1.1 Diffusion-limited regime

As shown in Figure 2.6, polyelectrolyte motion is dominated by diffusion in the absence of an electric potential and a free energy barrier. With $J(x) \sim nD$ and the Stokes−Einstein relation it follows that:

$$J \sim \frac{1}{R_g} \tag{2.24}$$

The radius of gyration R_g of the polymer can be expressed in terms of the number of monomers of the polymer and it has been shown that $R_g \sim 1/N^{0.6}$, where $\nu = 0.6$ is the Flory exponent for an uncharged polymer. Thus in the diffusion-limited regime, the flux can be described by [27]:

$$J \sim \frac{n}{N^{0.6}} \tag{2.25}$$

which implies that J decreases with increasing polymer length and increases with polymer concentration.

2.5.1.2 Drift regime

In the drift-dominated regime, the polymer moves only due to the applied external potential. This electrophoretic motion depends on the bare charge of the monomers and the salt concentration of the buffer. With increasing electrolyte concentration n the effective charge of the polymer decreases. As discussed before, this effect can be understood by considering the counterions shielding the surface charge. For a polyelectrolyte consisting of N-independent chains, the electrophoretic mobility μ can be described as follows [28]:

$$\mu = Ng(n) \tag{2.26}$$

where $g(n)$ is a function of salt concentration only that decreases with increasing n. Combining Eq. (2.26) with Eq. (2.23) and assuming a linear potential drop Ψ_0 leads to:

$$J(x) \sim nN\Psi_0 \tag{2.27}$$

Thus, in the drift-dominated regime, the polymer flux is linearly proportional to the electrolyte concentration n and the chain length N of the polyelectrolyte.

2.5.1.3 Free energy barrier

A translocation through a nanopore requires the localization of one free polymer end in front of the orifice followed by capturing. Therefore a successful translocation is accompanied by overcoming two free energy barriers denoted f_l and f_t, respectively. Muthukumar et al. showed that for polymers of sufficient length N that [27]:

$$f_l \sim \frac{1}{N^\alpha} \tag{2.28}$$

where $\alpha \approx 0.2 \pm 0.1$. Equation (2.28) implies that the free energy barrier associated with the localization of the free polymer end in front of the pore scales inversely with the number of monomers. This somewhat counterintuitive result can be understood by taking into account the entropic pressure at the pore entrance [27]. This pressure originates from the confining region required to push the end of the polymer in the direction of the pore entrance [24].

The second part of the free energy barrier f_t can be described as being directly proportional to the chain length of the polymer:

$$f_t \sim N \tag{2.29}$$

As Eqs (2.28) and (2.29) show opposite behavior for the dependency on the number of monomers N, the free energy barrier is often complex and difficult to describe analytically. For nanopores embedded in a thin membrane Muthukumar et al. showed that f_l dominates the energy barrier and thus $f_l \sim 1/N^\alpha$ [27].

2.5.2 Discussion of successful translocation

Equations (2.20) and (2.23) allow analysis of the capture rate and the probability of a successful translocation of a polymer across the nanopore. The latter requires the passage of the free energy maximum f_m which originates from the entropic barrier experienced by the chain. As discussed above, this barrier arises from the localization of a free polymer end at the orifice and the conformational changes required to initialize the translocation. Obviously, the structure of the free energy landscape can include several stable and metastable states and is shaped by the interaction between the pore and the polyelectrolyte [27].

In the following text, the general results will be summarized. For a given free energy landscape, the analysis of Eqs (2.20) and (2.23) reveals the existence of a threshold potential Ψ_m above which the flux J increases linearly. In contrast to this drift-dominated regime, the finite number of capture events below Ψ_m belongs to the diffusion-limited regime. Please note that with increasing chain length the barrier height decreases as indicated in Eq. (2.28).

The capture rate is a necessary but not sufficient criterion for the passage of a polymer across the nanopore. The probability of a successful translocation is again a direct consequence of the free energy landscape and the external voltage which drives the molecule over the energy barriers. Obviously, the higher the barrier the lower is the translocation rate. At a given potential, the probability of successful translocation is inversely proportional to the chain length N.

Briefly, this discussion shows that the capture rate and the probability of a successful translocation of a polymer across a nanopore depend on the height and location of the free energy barrier as well as the voltage difference. Three regimes can be distinguished. Below a certain threshold potential Ψ_m the polymer dynamics are governed by diffusion processes only. Above this critical value the polyelectrolyte is driven by the electric field in this drift-dominated regime. In both cases, the free energy landscape inside the nanopore is a key quantity in understanding translocation rates qualitatively and quantitatively.

It is important to notice that the presented theory is based on two assumptions. First, the polymer has to be able to explore all structural conformations while translocating the nanopore. Second, the polyelectrolytes enter the pore with one free end first. The experiments of the last decades, however, showed that DNA can enter the pore in single or multihairpin structures as well which is illustrated in Figure 2.7. Although these folding events are less likely to occur than the single entry, it would be important to integrate these facts into the theory. With a conclusive theoretical description, it might be possible to determine DNA stiffness or persistence length with a nanopore measurement. Of course, geometry will always be important for these measurements and thus a better control of nanopore geometry during fabrication would be desirable. One approach to this is the introduction of nanopores in graphene sheets [29–31].

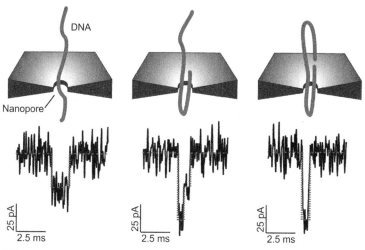

FIGURE 2.7

DNA substructures are indicated by their specific ionic current signature. Top: Sketches of linear, partly folded and circular DNA molecules passing through a solid-state nanopore. Below: single molecule traces indicating the corresponding ionic current change for the configurations shown above, as measured in a typical DNA translocation experiment.

Source: Adapted from Ref. [4].

2.6 Stalling DNA in a nanopore
2.6.1 Silicon nitride nanopore with optical tweezers

As we discussed, the use of the resistive pulse measurements on DNA using a nanopore allows for studies of the apparent (or effective) charge and the conformation of the translocating polyelectrolyte. Until now we discussed the passage of DNA based on the distribution of ions in cylindrical geometries using a linearized Poisson–Boltzmann model. Although this contains a number of simplifications, the description captures the essential ideas required to understand this complex process.

In order to gain deeper insight and perhaps a quantitative model of DNA translocation, more information about the translocation process is needed. While the free translocation of DNA allows determination of the time a molecule spends in the nanopore, it does not necessarily contain information about the velocity of the molecule. Several studies, in both larger [6] and smaller nanopores [32−34] have shown, that the translocation time can be a power law as a function of DNA contour length. These findings indicate that there might be other relevant processes. The origin of these power laws remains under intense investigation especially using computer-based simulations and experiments.

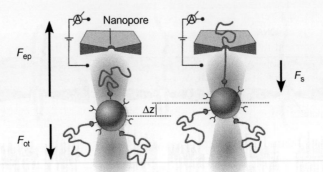

FIGURE 2.8

DNA tug of war. Electrophoretic force measurements based on optical tweezers. A colloidal particle (bead) coated with DNA is held in an optical trap in close proximity to a solid-state nanopore. The insertion of a DNA molecule into the nanopore results in a force on the bead indicated by a displacement Δz [36]. The arrows indicated the magnitude and direction of the electrophoretic force F_{ep} and the electroosmotic Stokes force F_s as well as the counteracting harmonic restoring force F_{ot} in the optical trap.

Source: Adapted from Ref. [37].

Extended structural and mechanical analyses of molecules in confinement require the incorporation of additional technologies, e.g., optical tweezers. After their introduction in 1986, optical tweezers became one of the workhorses in Biophysics [35]. Build around a strongly focused laser beam, optical tweezers allow single molecules to be manipulated with nanometer accuracy and without any mechanical contact. Often the molecule of interest, e.g. DNA, is attached to a colloidal probe which acts as a handle in the optical trap as shown in Figure 2.8[2].

Plenty of protocols are available to attach a finite number of DNA molecules onto a colloidal particle called bead in the following discussion. The majority involves functionalization of one end of double-stranded DNA, e.g., by biotin and subsequent incubation with streptavidin-modified beads. The biotin−streptavidin interaction is one of the strongest noncovalent bonds with a dissociation constant $K_A = 10^{-15}$.

In a typical optical tweezers experiment, a DNA-coated bead is located in front of a nanopore and an external potential is applied as indicated in Figure 2.8. Please note that all sketches in this section are not drawn to scale. The advantage of the optical trap is that the local concentration of DNA in front of the nanopore can be controlled by choosing the distance between bead surface and nanopore entrance. This unique ability of optical tweezers greatly facilitates the insertion and stalling of a single DNA molecule and still remains one of the key advantages over other combinations of nanopores with atomic force microscopy (AFM) [38] or magnetic tweezers [39].

When the DNA on the bead is being held within a capture radius r_c (see Section 2.3), a single molecule of DNA is eventually pulled inside the nanopore due to the electrophoretic force F_{ep}. This is indicated by a displacement Δz of the bead and ideally with a simultaneous change in ionic current. The measurement of both the force and the ionic current is key in being able to count the number of molecules in the nanopore. In the following text, we will assume that the force on the bead is independent of the finite but small electric field outside of the nanopore. It is also assumed that the electroosmotic flow along the DNA molecule and nanopore surfaces are irrelevant for all force measurements. This assumption is reasonable as most of the electric field is confined within the capture radius.

Utilizing the harmonic potential shape of the laser trap, this displacement can be related to a counteracting optical force F_{ot} on the colloid by a proportionality constant k_{trap}. The stalled DNA is also subjected to a Stokes friction force F_s due to the electroosmotic flow inside the nanopore which will be discussed below. In summary, the force balance can be described as follows:

$$F_{ep} = F_s + F_{ot} \tag{2.30}$$

The force analysis starts with the Poisson equation (Eq. (2.1)) in cylindrical coordinates and the Stokes equation (Eq. (2.15)). Combining both leads to:

$$\frac{1}{r}\frac{\partial}{\partial r}\left(r\frac{\partial v}{\partial r}\right) - \frac{\varepsilon}{\eta}\frac{1}{r}\frac{\partial}{\partial r}\left(r\frac{\partial \phi(r)}{\partial r}\right)E(x) = 0 \tag{2.31}$$

which has the general solution:

$$v(r) = \frac{\varepsilon}{\eta}E(x)\phi(r) + C_1 + C_2 \ln(a) \tag{2.32}$$

The constants C_1 and C_2 in Eq. (2.22) can be determined using the no-slip boundary condition at DNA and nanopore surface:

$$v(r_0) = 0$$

and

$$v(r_n) = 0 \tag{2.33}$$

which results in:

$$v(r) = \frac{\varepsilon}{\eta}E(x)\left(\phi(r) - \phi(r_0) + (\phi(r_0) - \phi(r_n))\frac{\ln((r/r_0))}{\ln((r_0/r_n))}\right) \tag{2.34}$$

Equation (2.32) describes the liquid velocity profile $v(r)$ inside a cylindrical nanopore in the presence of a single DNA molecule. It should be noted that the electrostatic potential $\phi(r)$ has to be obtained from the PBE as discussed in Section 2.1.

The electroosmotic flow inside the nanopore exerts a drag onto the DNA which is described by the shear stress:

$$\tau(r) = -\eta\frac{dv(r)}{dr} \tag{2.35}$$

Integrating the shear stress $\tau(r)$ along the length of the DNA inside the pore L allows for calculation of the drag force F_s on the molecule:

$$F_s = 2\pi r_0 L \tau(r_0)$$

$$F_s = 2\pi r_0 L \varepsilon E(x) \left(\frac{d\phi(r)}{dr} \Big|_{r=r_0} + \frac{(\phi(r_n) - \phi(r_0))}{r_0 \ln((r_0/r_n))} \right) \tag{2.36}$$

In order to derive the drag force on the polymer, the surface potential of both the DNA and the nanopore needs to be determined as indicated in Eqs (2.34) and (2.36). As $\phi(r_0)$ and $\phi(r_n)$ are not directly accessible, the zeta-potential ζ as introduced in the previous chapter can be applied. It should be mentioned here that the ζ potential, very much like the effective charge, is not to be regarded as an intrinsic material parameter in the context of nanopore measurements. It depends on the interaction of the DNA with its environment, depending on salt concentration. In other words, as an effective quantity ζ is defined as the potential at the outer Helmholtz layer of the EDL. This approach allows simplification of Eq. (2.30) to:

$$F_{ep} - F_s = F_{eff} = F_{ot} \tag{2.37}$$

Equation (2.37) implies that the force on the colloid in the optical trap F_{ot} is a measure for the effective force on the DNA. Considering the harmonic shape of the laser potential, the displacement of the bead allows for calculation of F_{eff}. While this approach appears to be straightforward, it has to be emphasized that the zeta-potential determined with a stalling force measurement is not a geometry independent quantity. As the zeta-potential is a function of salt concentration n_0, this result also implies that the DNA charge depends on the salt concentration and nanopore diameter [9,10,40]. A detailed discussion of the EDL and ζ can be found in the specialized literature and goes beyond the scope of this chapter.

2.7 Stalling DNA in nanocapillaries

Throughout this chapter we assumed that nanopores are easily available and did not discuss in detail the fabrication. This is done elsewhere in the book. For clarification of the physical effects of DNA translocation we discussed until now a combination of nanopores with optical tweezers as well as free translocation studies. These nanopores could be mainly described with a cylindrical geometry. In fact, almost all experiments described in the chapter until now, were based on the standard nanopores derived by drilling a single hole into a Si_3N_4 membrane using a focused electron beam of a transmission electron microscope [1]. Although this process is, without any doubt, the key for many of the great experiments described so far, it poses certain experimental challenges and is also quite cost and labor intensive. In the remaining paragraphs of this chapter, we will now describe an alternative source for nanopores, namely glass nanocapillaries.

Over the last decade glass capillaries have emerged as an alternative to solid-state nanopores made in plane membranes [41]. Using a laser pipette, puller capillaries are locally heated and torn apart. Depending on several parameters, such as material (borosilicate or quartz glass), capillary thickness, temperature, and pulling force, aperture diameters down to a few nanometers have been demonstrated [42]. Micro- and nanocapillaries are used for patch-clamp experiments [43] and for surface analysis experiments based on scanning ion conducting microscopy [44]. Recent experiments also showed the application of nanocapillaries to investigate the secondary structure and conformation of single double-stranded DNA (dsDNA) molecules of different length [45,46].

Figure 2.9A shows a scanning electron microscopy (SEM) image of a nanocapillary having a typical orifice diameter of 20 nm. In order to study the translocation of polyelectrolytes, the nanocapillary is embedded in a microfluidic chip where this constriction zone separates two reservoirs. Figure 2.9B displays an example of a microfluidic chip made of polydimethylsiloxane (PDMS).

2.7.1 Electrostatic characterization

In contrast to membrane-based nanopores, a nano- or microcapillary can be described by a cylindrical tip of length h_0 and diameter $2R_0$. Adjacent, the capillary opens with an opening angle α until the original inner diameter is reached

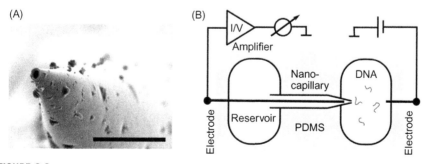

FIGURE 2.9

Microfluidic cell design and nanocapillary. (A) SEM image of a quartz nanocapillary. The scale bar represents 1 μm. The image is a courtesy of Lorenz Steinbock. (B) Microfluidic cell based on a PDMS spacer having two reservoirs connected by a nanocapillary only. A cover glass seals the fluidic cell. A potential across the nanocapillary is applied using Ag/AgCl electrodes. The current is measured using a standard amplifier for electrophysiology.

Source: Adapted from Refs [45,46].

which is shown in the upper part of Figure 2.10. The conical part of the capillary is referred to as the taper and has a typical length $L-h_0$ of several millimeters. As indicated in Figure 2.9B, the cylindrical rear part spans the gap between both reservoirs of the microfluidic cell.

The electrostatic properties of a nanocapillary can be analyzed using a finite element software, e.g., Comsol 4.1 (Comsol Multiphysics, Germany). This numerical approach separates the geometry into a grid of finite elements where the governing equations are being solved utilizing the appropriate boundary conditions. The lower part of Figure 2.10 shows a one-dimensional plot of the electrical potential distribution along the symmetry axis S calculated for a typical

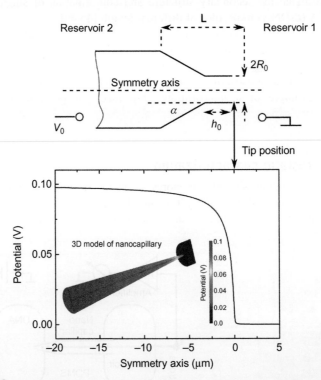

FIGURE 2.10

Geometry of a nanocapillary and electrostatic potential distribution. The potential is determined based on finite element calculations and plotted along the symmetry axis S of the nanocapillary. The tip is located at $S = 0\ \mu m$. The sketch on top shows a two-dimensional representation of the capillary tip region and highlights important parameters like the taper length L, the tip radius R_0, the tip length h_0, and the taper opening angle α. The axis of symmetry S is shown in the center of the capillary. The three-dimensional inset color codes the potential distribution inside the tip region of the glass capillary.

experimental configuration and with an applied potential of $V_0 = 100$ mV. The plot focuses around the capillary tip which is located at $S = 0$ μm. As expected the majority of the electric potential drops over the first micrometer and after 10 μm the potential is reduced to 90% accordingly.

The fact of a nonlinear potential drop over the sensing volume of a nanocapillary will be important when discussing force measurements on single double-stranded DNA molecules.

The tip diameter of the glass capillary can be derived by recording the current–voltage characteristic of the microfluidic cell. Adapting Eq. (2.18), the gradient of the current–voltage characteristic G can be used to estimate the tip diameter d of the glass capillary:

$$d = \frac{4Gl}{\pi\sigma_s d_b} \tag{2.38}$$

where L is the length of the taper, σ_s is the specific conductivity of the solution, and d_b is the full inner diameter of the capillary. The current–voltage characteristic shown in Figure 2.11 was measured in 20 mM KCl, 2 mM Tris pH 8 solution and revealed a conductance of $G = (9.50 \pm 0.05)$ nS. Utilizing values of $L = 1$ mm, $d_b = 0.325$ mm, and $\sigma_s = 2.64$ mS/cm, Eq. (2.37) yields an orifice diameter of around 150 nm. This is close to the typical diameter for these nanocapillaries of 140 nm obtained from SEM images.

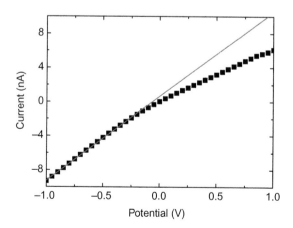

FIGURE 2.11

Electrostatic characterization of nanocapillary. Current–voltage characteristic of a 150 nm nanocapillary measured in 20 mM KCl and 2 mM Tris pH 8 solution. A linear fit to the negative part of the curve results in a conductance of (9.50 ± 0.05) nS. The rectification is due to the asymmetric shape of the capillary tip and the surface charge of the glass.

2.7.2 **Force measurements inside glass capillaries**

As in planar membranes, nanopores in glass capillaries can be used as a force sensor in combination with optical tweezers. In a typical experiment, λ-DNA coated polystyrene colloids are flushed into the microfluidic cell shown in Figure 2.9B and the optical trap is calibrated [47]. After calibration, a λ-DNA coated colloid is positioned in front of the capillary tip at a distance of several micrometers. Whilst applying a constant positive potential with respect to the capillary, the distance between colloid and tip is decreased using a piezoelectric nanopositioning system. The successful capture of a single double-stranded DNA molecule is indicated by a change in force.

As discussed in Section 3.1, the optical tweezers act as a sensor for the effective force on the DNA. Figure 2.12 shows a typical data trace of measurements on a single λ-DNA molecule inside a nanocapillary.

Within the first 25 s, no net force on the colloid is visible. At $t = 25$ s the force decreases by approximately 7 pN in a single step. In order to probe the electrostatic properties, the DNA is slowly pulled out of the nanocapillary using a piezoelectric nanopositioning system. In Figure 2.12 pulling starts at approximately 30 s. Due to the nonlinear potential drop across the sensing region, the magnitude of the force decreases slowly while the DNA is pulled out. In contrast to nanopores made of planar membranes, this property of nanocapillary allows them to be used as a counter of the effective DNA length inside the pore. When the last part exits the tip the force jumps to 0 as indicated in Figure 2.12. This behavior can also be understood using an adapted analytical model from Ghosal [48]. Based on simple geometrical considerations, it is possible to calculate the force on the DNA as a function of colloid-tip distance. A comparison between experiment and model shows a very good quantitative agreement.

These experiments and modeling results show that, with the physical description of the process of DNA translocation, a quantitative understanding is possible. This is remarkable taking into account the number of assumptions and simplifications in the model. However, as discussed in the beginning, the basic physical principles should hold for both static DNA molecules in nanopores or moving DNA. Although there are several caveats preventing a direct comparison of the results from the optical tweezers—based force measurements with the forces encountered during the dynamic translocation process, the main conclusion still stands that the underlying principles can be well understood with a description of the problem with the PBE.

It should be mentioned here that approaches using the Poisson—Nernst—Planck (PNP) equation are also very successful in interpreting the data of nanopore experiments. However, we believe that the PBE allows a more intuitive understanding, especially in the linearized form. The PNP equation is much more difficult to treat analytically and can be only solved numerically, which prevents a clear insight into the basic principles.

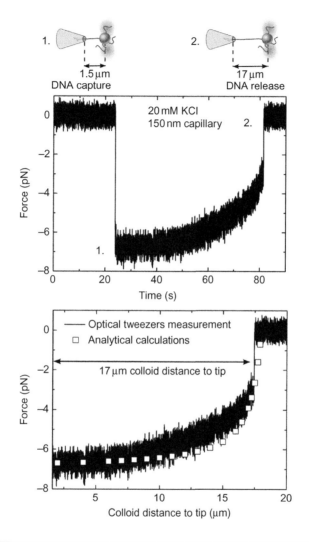

FIGURE 2.12

The upper graph shows λ-DNA capture inside a glass capillary indicated by a step in optical force at $t = 25$ s. Using a piezoelectric nanopositioning system, the colloid is moved away from the capillary tip and the λ-DNA is eventually pulled out at $t = 85$ s at a distance close to its contour length. Pulling starts at approximately 30 s. The lower graph shows the force on the DNA as a function of colloid-tip distance. The scatter graph (□) is obtained using an analytical model.

Source: Adapted from Ref. [48].

References

[1] Dekker C. Solid-state nanopores. Nat Nanotechnol 2007;2:209−15. doi:10.1038/nnano.2007.27.

[2] Keyser UF. Controlling molecular transport through nanopores. J R Soc Interface 2011;8:1369−78. doi:10.1098/rsif.2011.0222.

[3] Li JL, Gershow M, Stein D, Brandin E, Golovchenko JA. DNA molecules and configurations in a solid-state nanopore microscope. Nat Mater 2003;2:611−5. doi:10.1038/Nmat965.

[4] Smeets RMM, Keyser UF, Krapf D, Wu MY, Dekker NH, Dekker C. Salt dependence of ion transport and DNA translocation through solid-state nanopores. Nano Lett 2006;6:89−95. doi:10.1021/nl052107w.

[5] Storm AJ, Chen JH, Zandbergen HW, Dekker C. Translocation of double-strand DNA through a silicon oxide nanopore. Phys Rev E 2005;71:10.1103/Physreve.71.051903.

[6] Storm AJ, Chen JH, Ling XS, Zandbergen HW, Dekker C. Fast DNA translocation through a solid-state nanopore. Nano Lett 2005;5:1193−7. doi:10.1021/Nl048030d.

[7] Wanunu M, Sutin J, McNally B, Chow A, Meller A. DNA translocation governed by interactions with solid-state nanopores. Biophys J 2008;95:4716−25. doi:10.1529/biophysj.108.140475.

[8] Delgado AV, Gonzalez-Caballero F, Hunter RJ, Koopal LK, Lyklema J. Measurement and interpretation of electrokinetic phenomena. J Colloid Interf Sci 2007;309:194−224. doi:10.1016/j.jcis.2006.12.075.

[9] Keyser UF, van Dorp S, Lemay SG. Tether forces in DNA electrophoresis. Chem Soc Rev 2010;39:939−47. doi:10.1039/b902072c.

[10] van Dorp S, Keyser UF, Dekker NH, Dekker C, Lemay SG. Origin of the electrophoretic force on DNA in solid-state nanopores. Nat Phys 2009;5:347−51. doi:10.1038/nphys1230.

[11] Hunter RJ. Zeta potential in colloid science: principles and applications. London: Academic Press; 1981.

[12] Kowalczyk SW, Grosberg AY, Rabin Y, Dekker C. Modeling the conductance and DNA blockade of solid-state nanopores. Nanotechnology 2011;22:10.1088/0957-4484/22/31/315101.

[13] Poon WCK, Andelman D. Soft condensed matter physics in molecular and cell biology. New York; London: Taylor & Francis; 2006.

[14] Stein D, Kruithof M, Dekker C. Surface-charge-governed ion transport in nanofluidic channels. Phys Rev Lett 2004;93:10.1103/Physrevlett.93.035901.

[15] Yusko EC, Johnson JM, Majd S, Pragkio P, Rollings RC, Li J, et al. Nanopores with fluid walls. Biophys J 2011;100:169.

[16] Hernandez-Ainsa S, Muss C, Bell NAW, Steinbock LJ, Thacker VV, Keyser UF. Lipid-coated nanocapillaries for DNA sensing. Analyst 2013;138:104−6.

[17] Firnkes M, Pedone D, Knezevic J, Doblinger M, Rant U. Electrically facilitated translocations of proteins through silicon nitride nanopores: conjoint and competitive action of diffusion, electrophoresis, and electroosmosis. Nano Lett 2010;10:2162−7. doi:10.1021/nl100861c.

[18] Chang H, Kosari F, Andreadakis G, Alam MA, Vasmatzis G, Bashir R. DNA-mediated fluctuations in ionic current through silicon oxide nanopore channels. Nano Lett 2004;4:1551−6.

[19] Fan R, Karnik R, Yue M, Li D, Majumdar A, Yang P. DNA translocation in inorganic nanotubes. Nano Lett 2005;:10.1021/nl0509677.

[20] Bezrukov SM. Ion channels as molecular coulter counters to probe metabolite transport. J Membr Biol 2000;174:1−13.

[21] Gracheva ME, Melnikov DV, Leburton JP. Multilayered semiconductor membranes for nanopore ionic conductance modulation. ACS Nano 2008;2:2349−55. doi:10.1021/Nn8004679.

[22] Timp W, Mirsaidov UM, Wang D, Comer J, Aksimentiev A, Timp G. Nanopore sequencing: electrical measurements of the code of life. IEEE Trans Nanotechnol 2010;9:281−94. doi:10.1109/Tnano.2010.2044418.

[23] Vlassiouk I, Smirnov S, Siwy Z. Nanofluidic ionic diodes. Comparison of analytical and numerical solutions. ACS Nano 2008;2:1589−602. doi:10.1021/Nn800306u.

[24] Muthukumar M. Polymer translocation. Boca Raton; London; New York: CRC Press; 2011.

[25] Howorka S, Siwy Z. Nanopore analytics: sensing of single molecules. Chem Soc Rev 2009;38:2360−84. doi:10.1039/b813796j.

[26] Wanunu M, Morrison W, Rabin Y, Grosberg AY, Meller A. Electrostatic focusing of unlabelled DNA into nanoscale pores using a salt gradient. Nat Nanotechnol 2010;5:160−5. doi:10.1038/Nnano.2009.379.

[27] Muthukumar M. Theory of capture rate in polymer translocation. J Chem Phys 2010;132:10.1063/1.3429882.

[28] Muthukumar M. Mechanism of DNA transport through pores. Annu Rev Biophy Biomol Struct 2007;36:435−50. doi:10.1146/annurev.biophys.36.040306.132622.

[29] Garaj S, Hubbard W, Reina A, Kong J, Branton D, Golovchenko JA. Graphene as a subnanometre trans-electrode membrane. Nature 2010;467:190−3. doi:10.1038/nature09379.

[30] Merchant CA, Healy K, Wanunu M, Ray V, Peterman N, Bartel J, et al. DNA translocation through graphene nanopores. Nano Lett 2010;10:2915−21. doi:10.1021/Nl1010461.

[31] Schneider GF, Kowalczyk SW, Calado VE, Pandraud G, Zandbergen HW, Vandersypen LMK, et al. DNA translocation through graphene nanopores. Nano Lett 2010;10:3163−7. doi:10.1021/nl102069z.

[32] Li JL, Talaga DS. The distribution of DNA translocation times in solid-state nanopores. J Phys Condens Matter 2010;22:10.1088/0953-8984/22/45/454129.

[33] Niedzwiecki DJ, Grazul J, Movileanu L. Single-molecule observation of protein adsorption onto an inorganic surface. J Am Chem Soc 2010;132:10816−22. doi:10.1021/Ja1026858.

[34] Lu B, Albertorio F, Hoogerheide DP, Golovchenko JA. Origins and consequences of velocity fluctuations during DNA passage through a nanopore. Biophys J 2011;101:70−9. doi:10.1016/j.bpj.2011.05.034.

[35] Neuman K, Block S. Optical trapping. Rev Sci Instrum 2004;75:2787−809.

[36] Keyser UF, van der Does J, Dekker C, Dekker NH. Optical tweezers for force measurements on DNA in nanopores. Rev Sci Instrum 2006;77:10.1063/1.2358705.

[37] Keyser UF, Koeleman BN, van Dorp S, Krapf D, Smeets RMM, Lemay SG, et al. Direct force measurements on DNA in a solid-state nanopore. Nat Phys 2006;2:473−7. doi:10.1038/nphys344.

[38] King GM, Golovchenko JA. Probing nanotube−nanopore interactions. Phys Rev Lett 2005;95:10.1103/PhysRevLett.95.216103.

[39] Peng HB, Ling XSS. Reverse DNA translocation through a solid-state nanopore by magnetic tweezers. Nanotechnology 2009;20:10.1088/0957-4484/20/18/185101.

[40] Hall AR, van Dorp S, Lemay SG, Dekker C. Electrophoretic force on a protein-coated DNA molecule in a solid-state nanopore. Nano Lett 2009;9:4441−5. doi:10.1021/nl9027318.

[41] Lieberman K, Lewis A, Fish G, Shalom S, Jovin TM, Achim S, et al. Multifunctional, micropipette based force cantilevers for scanned probe microscopy. Appl Phys Lett 1994;65:648−50. doi:10.1063/1.112259.

[42] Bruckbauer A, Ying L, Rothery AM, Zhou D, Shevchuk AI, Abell C, et al. Writing with DNA and protein using a nanopipette for controlled delivery. J Am Chem Soc 2002;124:8810−1 ja026816c [pii].

[43] Neher E, Sakmann B. Single-channel currents recorded from membrane of denervated frog muscle fibers. Nature 1976;260:799−802. doi:10.1038/260799a0.

[44] Korchev YE, Milovanovic M, Bashford CL, Bennett DC, Sviderskaya EV, Vodyanoy I, et al. Specialized scanning ion-conductance microscope for imaging of living cells. J Microsc 1997;188:17−23. doi:10.1046/j.1365-2818.1997.2430801.x.

[45] Steinbock LJ, Otto O, Chimerel C, Gornall J, Keyser UF. Detecting DNA folding with nanocapillaries. Nano Lett 2010;10:2493−7. doi:10.1021/Nl100997s.

[46] Steinbock LJ, Otto O, Skarstam DR, Jahn S, Chimerel C, Gornall JL, et al. Probing DNA with micro- and nanocapillaries and optical tweezers. J Phys Condens Matter 2010;22:10.1088/0953-8984/22/45/454113.

[47] Tolic-Norrelykke SF, Schaffer E, Howard J, Pavone FS, Julicher F, Flyvbjerg H. Calibration of optical tweezers with positional detection in the back focal plane. Rev Sci Instrum 2006;77:10.1063/1.2356852.

[48] Ghosal S. Electrophoresis of a polyelectrolyte through a nanopore. Phys Rev E 2006;74:10.1103/Physreve.74.041901.

Instrumentation for Low-Noise High-Bandwidth Nanopore Recording

3

Vincent Tabard-Cossa

Center for Interdisciplinary NanoPhysics, Department of Physics, University of Ottawa,
150 Louis Pasteur, Ottawa, ON, K1N 6N5, Canada

CHAPTER OUTLINE

Engineered Nanopores for Bioanalytical Applications.
© 2013 Elsevier Inc. All rights reserved.

3.1 Introduction

Nanopores have become a versatile single-molecule analytical tool for high-throughput, label-free characterization of individual nucleic acids and protein molecules [1−5]. The method relies on translating the properties or the activity of an individual biomolecule into an electrical signal by momentarily confining it in the nanoscale geometry of a pore [6−8]. When a small voltage bias (∼0.1−1 V) is applied across a nanometer-sized pore embedded in a thin insulating membrane separating two reservoirs of liquid electrolyte (typically 0.01−1 M KCl), the resulting ionic current through the pore (∼0.01−100 nA) can be measured with standard low-current recording techniques. As a biomolecule is electrophoretically driven through a nanopore, the resulting transient change in the ionic current indicates the presence of the analyte. The properties of these electrical pulses (e.g., shape and duration) can provide structural information (e.g., length, size, charge, and shape) about the molecule of interest [9−14] and/or may be used to characterize biomolecular binding and interaction strengths [15−18].

Biological pores were first used to detect individual poly(ethylene glycol) polymers in 1994 [19] and ssDNA molecules in 1996 [20]. Since that time, a host of applications for nucleic acid and protein analysis using both biological and solid-state nanopores have been developed [2,5]. Each type of pore has distinct advantages and limitations [21]. For instance, biological pores offer atomically precise structure together with the engineering capability of site-directed mutagenesis. This, in general, translates into highly reproducible and stable ionic current flow through pores, which can be specifically designed to alter biomolecular passage speed and improve their chemical specificity [22,23]. Yet, the fragility of the supporting lipid bilayer and the fixed size of the pores restrict their applications. In contrast, while reproducible fabrication of low-noise solid-state nanopores with precise dimensions remains challenging [4], they offer increased durability over a wide range of operating conditions, a tuneable pore geometry, and a more natural propensity for integration with microfluidics and CMOS technologies for precise, rapid fluidic and electronic control [24,25], and mass-manufacturing of devices.

Noise and bandwidth of ionic current recordings in both types of nanopores limit their sensitivity and reliability as single-molecule sensors because of the transient nature of the signals. This is particularly true of silicon nitride nanopores, which exhibit significantly larger noise levels and shorter-lasting events than biological pores, such as α-hemolysin [26]. This is mainly due to distinct geometries and material compositions between organic and synthetic devices, giving rise to a different noise response, and biomolecule-pore wall interactions. Recently, many solid-state nanopore studies have focused on analysis of proteins, or short DNA/RNA fragments [12,27−35], reporting short-lived events, which are generally difficult to detect because ionic current noise steeply rises with increasing bandwidth. Low signal levels, and fast translocation times are also a major barrier in the development of nanopore-based DNA sequencing applications [4,36−37]. Advancement in the field of nanopore-based sensing is therefore tied to an increase in sensitivity and accuracy of the electrical signal to further enhance the information content of the ionic current pulses. This involves reducing the level of electrical noise and improving temporal resolution of ionic current recordings. As discussed in this chapter, this is primarily achieved through the development of specially designed nanopore chips, and integrated amplifiers optimized for nanopore-sensing applications.

In this chapter, we introduce the necessary instrumentation for performing low-noise nanopore experiments, discuss the various sources of noise in detail, and review the progress made to date to improve signal-to-noise ratio (SNR) and enhance bandwidth in ionic current recordings. Admittedly, the chapter will emphasize solid-state nanopore devices over their organic counterparts, though due to their strong overlap much of the material presented is directly applicable to the case of biological nanopores.

Although conventional patch-clamp amplifiers, employed in electrophysiology, can be used to measure current with pA sensitivity in a bandwidth reaching 100 kHz, attention should be paid to the design of the entire nanopore recording setup, and the contributing noise sources, in order to realize such performance. As discussed in Section 3.2, this includes the architecture of the chip supporting the solid-state membrane or the lipid bilayer, the design of the fluidic cell that mounts nanopores into liquids, and their connection to computer-controlled current recording instruments. In Section 3.3, we review the operating principle of low-noise current amplifiers universally used to apply voltages across a membrane and measure the resulting ionic current through a nanopore. Knowledge of the electronic circuitry allows us to identify, in Section 3.4, the contributing sources of noise, including those associated with the nanopore, the measurement electronics, and those arising from their integration. This detailed review of signal bandwidth and noise sources provides general guidelines for optimizing nanopore instrumentation and achieving reliable low-noise recordings employing today's state-of-the-art commercially available current amplifiers. In Section 3.5 we examine noise generated by the digitizer, the use of electronic filters to suppress high-frequency noise, and discuss sampling considerations to maximize temporal resolution and SNR. We conclude this chapter with an outlook (Section 3.6) reviewing the recent progress in low-noise measurement platforms explicitly designed for nanopore sensing of individual nucleic acid and protein molecules.

3.2 Components of a nanopore setup and their integration

A complete nanopore recording instrument integrates multiple building blocks. The guidelines for assembling a nanopore setup optimized for low-noise, high-bandwidth ionic current measurements are comparable to those established for patch-clamp recordings employed in electrophysiology [38−41]. As such the Axon Guide [42] is a very useful reference. Figure 3.1 depicts the schematic diagram of a commonly configured nanopore recording setup.

A nanopore is mounted in a polytetrafluoroethylene (PTFE) fluidic cell designed to form a fluidic-tight seal between two liquid electrolyte-filled reservoirs on either side of the membrane. The cell must ensure that the nanopore is the only liquid junction between the two reservoirs. Ag/AgCl electrodes, traditionally used to apply the voltage across the nanopore and measure the resulting ionic current, are immersed on either side of the membrane. The fluidic cell is embedded in a tight metallic enclosure to provide electromagnetic (EM) shielding and to securely fix the electrodes. The amplifier headstage connects to the electrodes in close proximity to the fluidic cell enclosure. A secondary Faraday cage made from solid folded copper sheet can also be used to provide additional EM shielding to the assembly. The setup is positioned on an isolation table to minimize noise induced by mechanical vibrations, which are detrimental to the lipid bilayer integrity, and can act as a source of electrical noise through vibrations of the electrodes. Temperature regulation with ~0.1°C precision can be achieved by using a water-cooled thermoelectric element positioned underneath the secondary shielding box. The analog signals of the current amplifier are digitized with a computer-controlled data acquisition (DAQ) card. The DAQ card is also used to control the command voltage of the amplifier, which sets the electric potential across the membrane. The DAQ card should have the following minimal requirements: 2 analog inputs, 1 analog output, 250 kHz sampling rate per channel, 16-bit resolution, and ±10 V dynamic range. In parallel, the user can decide to connect an

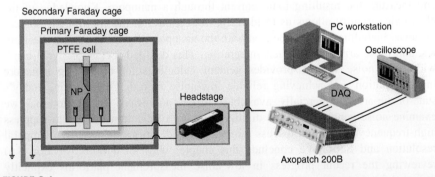

FIGURE 3.1

Schematic diagram of a complete nanopore recording setup. The nanopore is mounted in a fluidic cell made of PTFE, and separates two liquid reservoirs. The design of the cell must ensure that the nanopore is the only liquid junction between the two reservoirs. The fluidic cell is embedded in primary Faraday cage to shield from electromagnetic interferences. Ag/AgCl electrodes immersed on either side of the nanopore are connected to the remote headstage of an instrument sourcing voltage and measuring current (e.g., Axopatch 200B from Molecular Devices). A secondary Faraday cage made from a folded sheet of copper provides additional shielding to the fluidic cell/headstage assembly. The analog signals of the amplifier are digitized with a DAQ card connected to a computer. The DAQ card is also used to control the amplifier's sourced voltage. An oscilloscope can be used to monitor the signals in real time.

oscilloscope at the outputs of the current amplifier for real-time monitoring of the ionic current signals and applied voltages. The sections below discuss special considerations for designing and integrating each individual component.

3.2.1 **Nanopore support structure**

Biological and solid-state nanopores are formed in very different ways. α-Hemolysin, for instance, self-organizes into a lipid bilayer membrane [43], while silicon nitride nanopores are generally drilled via a transmission electron microscope (TEM) in a freestanding membrane [44,45]. Despite this obvious difference, both systems require a solid support for their membrane. This support structure can come in the form of a small aperture fabricated at the end of a PTFE tube for biological pores, and as a silicon-based substrate for solid-state devices. Figure 3.2A depicts the schematic of a proteinaceous pore incorporated in a lipid bilayer suspended across a small aperture. Figure 3.2B shows the device schematic of a solid-state nanopore embedded in a thin silicon nitride membrane supported by a silicon substrate.

The silicon chip is usually 3 mm in size and 200 µm thick with a pyramidal pit defined by KOH etch to create a freestanding membrane in the center. Silicon nitride is currently the material of choice for forming robust and stable solid-state

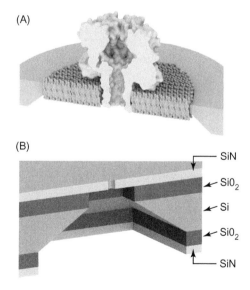

(A)

(B)

SiN

SiO$_2$

Si

SiO$_2$

SiN

FIGURE 3.2

(A) Schematic of a biological pore inserted into a lipid bilayer suspended across a small aperture typically 10–20 µm in diameter. (B) Schematic of a solid-state nanopore embedded in a thin (ca. 30 nm) insulating 50×50 µm^2 membrane supported by a 200 µm-thick silicon chip 3 mm in size. To reduce capacitance of the device, the silicon nitride membrane can be supported by a silicon dioxide layer a few microns thick.

nanopores, though other materials are being actively investigated [46–52]. The silicon nitride membranes formed in this way normally have ~50 μm lateral dimensions. To reduce the noise due to capacitance of the membrane, and to improve its mechanical strength, membrane windows should in principle be fabricated as small as possible. However, it is difficult to achieve smaller feature sizes with high yield using standard MEMS fabrication techniques, due to the variability in the thickness of the wafer. Membrane thickness used for nanopore experiments is typically 20–50 nm. Thinner membranes are favored because the electrical pulse height from a single-molecule event is inversely proportional to the membrane thickness, t (i.e., $\Delta I \propto 1/t$) [53]. However, very thin membranes (~10 nm) tend to be more fragile and consequently reduce their yield. As illustrated in Figure 3.2B, more recent solid-state nanopore chip designs include a ~1–5 μm-thick silicon dioxide layer to insulate the silicon nitride membrane from the doped silicon substrate, in an effort to reduce capacitance of the nanopore support structure [54]. An overview of the progress made to date to design improved nanopore chips optimized for sensing applications is presented in Section 3.6.

Equivalent circuit models can be used to analyze electrical noise and the frequency response of a system. Following the detailed analysis of the equivalent circuit of several different architectures of solid-state nanopore devices [55–57], a generic equivalent circuit for a nanopore chip is presented in Figure 3.3. The electrical access resistance from the electrode/electrolyte interface to the membrane is accounted for by $R_{electrolyte}$, which in part depends on the design of the fluidic cell. R_{pore} represents the pore resistance, which is assumed to include

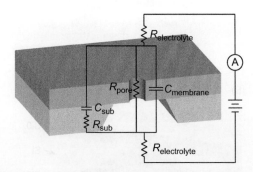

FIGURE 3.3

Simplified equivalent circuit superimposed on a nanopore schematic used to analyze electrical noise and the frequency response of the system. $R_{electrolyte}$ is the access resistance from the electrode interface to the membrane. R_{pore} is the nanopore resistance. R_{sub} is the electrical path through the doped silicon substrate and the dielectric layer(s). $C_{membrane}$ is the capacitance of the freestanding membrane. C_{sub} is the generalized capacitance through the substrate. It is composed of the silicon nitride layer capacitance in series with the capacitance of the silicon depletion layer at the silicon-electrolyte interface and the electrostatic double layer in solution.

Source: Adapted from Ref. [56].

effects from its access resistance and wall surface charge density [58−60]. R_{sub} is the electrical path through the doped silicon substrate and the dielectric layer(s). $C_{membrane}$ is the capacitance of the freestanding membrane. C_{sub} is the generalized capacitance through the substrate, composed of the silicon nitride layer capacitance in series with the capacitance of the silicon depletion layer at the silicon-electrolyte interface and the electrostatic double layer in solution [55,56]. It is dominated by the smallest capacitance in the path, which will most likely be either the capacitance associated with the silicon nitride layer or the silicon depletion layer. The generalized capacitance of the chip, C_{chip}, can be modeled by $C_{chip} = \sum_i \epsilon_0 \epsilon_{r_i} A_i / t_i$, where ϵ_0 is the permittivity of free space, ϵ_{r_i} the relative permittivity of the different dielectric layers forming the membrane and support structure, A_i is the area in contact with the electrolyte, and t_i is the thickness of the particular dielectric layers [25]. Modeling the nanopore system as a network of resistive and capacitive components allows us to estimate the maximum signal bandwidth of the nanopore sensor [55−57,61]. The signal bandwidth is set by the maximum cutoff frequency defined by the RC time constant of the nanopore support structure:

$$f_c \leq \frac{1}{2\pi RC} \tag{3.1}$$

where, in the particular case depicted in Figure 3.3, $R = 2R_{electrolyte}$, and $C = C_{chip} = C_{membrane} + C_{sub}$. In general, for standard solid-state nanopore chip and fluidic cell design, $R_{electrolyte} \sim 0.1−1\ k\Omega$, and $C_{chip} \sim 0.1−1\ nF$. Using these estimates one can expect a maximal signal bandwidth in the MHz range. As we will see in Section 3.3 this is significantly greater than the measurement bandwidth of the recording electronics ($\sim$100 kHz), and therefore does not impose a limit on the nanopore recording instrumentation. Nevertheless, capacitance of the nanopore device should ideally be minimized to the 1−10 pF range to reduce ionic current noise at high-frequency and significantly improve SNR at high-bandwidth measurements (see Section 3.4).

3.2.2 Fluidic cell

A fluidic cell securely mounts the nanopore in a liquid electrolyte to perform sensing experiments. The cell should be constructed in a chemically inert and resistant material (e.g., PTFE and polyether ether ketone (PEEK)), to permit thorough cleaning and decontamination in between experiments (Box 3.1). The material should also be insulating, with good dielectric properties as to not interfere with the electrical measurements.

As mentioned previously, a key feature of the nanopore fluidic cell is to form a liquid-tight seal between the two sides of the membrane, to ensure that the nanopore is the only fluidic junction across the two electrolyte-filled reservoirs. Multiple approaches have been proposed for forming a high-quality seal. In general, a pair of custom-fabricated gaskets is used, whereby the two halves of the fluidic cell compress the gaskets/nanopore chip assembly. Note that O-rings of

BOX 3.1 PROTOCOL FOR PERFORMING NANOPORE EXPERIMENTS

Steps involved in mounting a solid-state nanopore into liquids, as well as conditioning and characterizing it prior to performing single-molecule experiments include:

Cleaning and Mounting

1. Clean the fluidic cell by placing in a glass beaker a 5:1 ratio of deionized H_2O and concentrated HNO_3 and boiling it for 10 min. Rinse with copious amounts of filtered, deionized H_2O and dry under filtered air or dry N_2. Store in a clean dust-free environment until use. Alternatively, store in filtered H_2O until use.
2. Preset hot plate to 85°C in a fume hood.
3. Prepare piranha solution by pipetting 3 mL of concentrated H_2SO_4 followed by 1 mL of H_2O_2 into a 10 mL glass beaker. Gently reflux, with a disposable glass pipette, to ensure proper mixing of the solution. Piranha will rapidly heat up and may begin to boil. [Warning: piranha solution reacts strongly with organic compounds and should be handled with extreme caution; do not store the solution in closed containers, and dispose following hazardous waste disposal procedure of your institution.]
4. Using acid-resistant tweezers, place a nanopore chip in the piranha solution. To prevent floating and ensure complete immersion, insert chip vertically until it is completely submerged.
5. Rinse tweezers with copious amounts of H_2O to remove any trace of piranha solution.
6. Place the beaker containing the nanopore chip on the preheated hot plate for 30–45 min. Periodically check to ensure that the nanopore chip remains submerged, as bubbles can form on its surface and cause it to float.
7. While the nanopore chip is cleaning, degas ∼50 mL of filtered deionized H_2O and buffered salt solution. This can be done by sonicating in a heated water bath at 35°C followed by placement in a vacuum chamber for ∼10 min. Additionally, clean two silicone elastomer gaskets by placing in a small beaker containing 1:1 mixture of EtOH:H_2O and sonicating for 15 min. When clean, remove the gaskets from the solution, suction dry using an aspirator and place on a clean glass slide.
8. Remove the beaker containing the nanopore chip from the hot plate. Being careful not to damage the membrane, remove piranha solution from the beaker using a disposable glass pipette and discard in a waste beaker containing a large amount of H_2O.
9. Rinse the beaker and nanopore chip with ∼5 mL degassed filtered deionized H_2O, using the disposable glass pipette. Gently reflux and remove H_2O. Repeat 4–5 times.
10. Remove H_2O from the beaker and replace with methanol or ethanol.
11. Remove ethanol from the beaker. Using clean sharp-tip tweezers, gently grip the edge of the nanopore chip and remove it from the beaker.
12. Carefully dry the nanopore chip by applying suction to its edge (i.e., horizontally) using an aspirator. When dry, place the nanopore chip on a clean gasket such that the membrane window is aligned with the gasket opening.
13. Place the second gasket above the nanopore chip, again aligning the membrane window with the gasket opening. Inspect the chip under stereomicroscope at 10 × magnification equipped with coaxial episcopic illumination to verify membrane integrity and ensure proper alignment of the gaskets.
14. Place nanopore chip and gaskets into the fluidic cell and seal by firmly screwing the two half-cells together, see Figure 3.4.
15. To wet the nanopore, flush both reservoirs with methanol or ethanol. To remove any trapped air, place the cell in a vacuum chamber until trapped bubbles are observed to exit the fluidic cell channels.

16. Remove methanol or ethanol by flushing both reservoirs with 5 mL of degassed, buffered salt solution, drawing away overflow with an aspirator. Repeat this 2–3 times to completely remove alcohol.

Conditioning and Characterization

1. Place fluidic cell in a setup (see Figure 3.1) and insert Ag/AgCl electrodes into both electrolyte reservoirs. Connect the electrodes to the current amplifier headstage.
2. Apply 200 mV and observe the ionic current as a function of time. If the conductance is lower than expected or showing signs of significant low-frequency fluctuation (see Section 3.4), the nanopore may not be completely wetted or may be partially obstructed. In this case, disconnect the electrodes from the headstage and connect them to an external DC power supply. Apply short 100–200 ms pulses of 5–10 V amplitude while monitoring the ionic current at 200 mV in-between pulses until a stable, low-noise conductance is observed. Applying such a moderately high potential difference across the membrane will help completely wet the pore and/or remove any parasitic debris, which give rise to low-frequency current fluctuations. [Warning: applying such voltage pulses for an extended period of time can significantly enlarge the nanopore size, as discussed in Beamish et al. [62].]
3. Record $\sim$30 s of ionic current data at 200 mV and perform a PSD analysis of the current noise. If the I_{RMS} noise at 5 kHz is above 15 pA RMS, the nanopore is likely still not completely wetted or is partially clogged. If this is the case, continue to apply short pulses of 5–10 V using the external power supply and monitor the current between pulses at 200 mV.
4. Observe the I–V characteristics of the nanopore by ramping the voltage from −200 to +200 mV while recording the ionic current. For symmetric, wetted and clean nanopores, the I–V curve should be linear in this potential window. The slope of this curve provides a measure of conductance, from which the size of the nanopore under experimental conditions can be determined. If a particular application requires a specific nanopore size, the nanopore can be enlarged by applying subsequent pulses of 8–10 V until the desired conductance is reached as described in Ref. [62].

Unmounting

1. To remove the nanopore from the cell, first flush the reservoirs with H_2O to remove salt from the system.
2. Remove the nanopore chip and gaskets from the cell and place in ethanol. After $\sim$5 min, the gaskets will swell and can be removed from the nanopore chip using tweezers.
3. Rinse the nanopore chip with ethanol and dry by applying suction to its edge. It is now ready to use again following piranha cleaning as in steps 1–10.

the appropriate size do not offer a large contact surface area, and their aspect ratio can make it arduous to wet the nanopore. Gaskets can be microfabricated in poly-dimethylsiloxane (PDMS) to have a small inner diameter funnel-shaped aperture [53,54]. Exposing an area of the membrane to the electrolyte a few hundred microns in diameter, or less, helps reduce the chip capacitance, and associated high-frequency current noise [26,55]. However, adequately aligning such gaskets over the chip can be challenging. Sealing the nanopore chip can also be reliably achieved by painting a thin coat of a fast, room temperature curing silicone elastomer sealant directly on the chip [63]. While forming a high-quality seal,

(A)   (B)

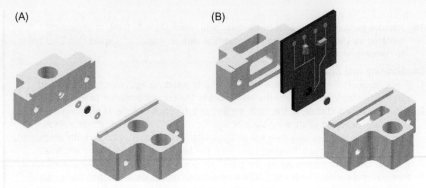

FIGURE 3.4

(A) Exploded view of a solid-state nanopore fluidic cell assembly. The nanopore chip is sandwiched in between two silicone elastomer gaskets to form a gigaohm-range seal between the two liquid reservoirs, where the electrodes are immersed. The extra reservoir can be filled with water to slow evaporation, or to calibrate temperature inside the cell with the measurement from a thermistor positioned outside the Faraday cage. (B) Exploded view of an electrically gated nanopore fluidic cell assembly. The gated nanopore is mounted on a custom-machined PCB to make electrical contact via a wirebond to the gate electrode. PDMS is used to seal the gated nanopore chip on the PCB, and insulate electrical contact. The PCB is then clamped between two half-cells with custom-made silicone gaskets to form a fluid tight seal, between the reservoirs.

unmounting the nanopore chip from the fluidic cell requires special solvents, and may leave some residues. Alternatively, custom gaskets can also be punched through silicone elastomer sheets [26]. This represent a cost-effective, and efficient way to form high-quality seals. Figure 3.4 shows examples of fluidic cells used to mount solid-state nanopores and electrically gated nanopores into liquids. The standard nanopore chip is simply sandwiched between two silicone elastomer gaskets by two PTFE half-cells. The nanopore equipped with an embedded metal electrode is mounted and sealed on a printed circuit board (PCB) with PDMS or a fast curing sealant. The PCB is then clamped between two half-cells with custom-made silicone elastomer gaskets.

Other important features of the fluidic cell include the volume of the reservoirs on each side of the membrane and the geometry of the fluidic access to the nanopore chip. For practical reasons (including immersion of Ag/AgCl electrodes) and to avoid issues arising from evaporation or electrolyte conductivity changes when introducing biological samples, volumes in the range of ~500 μL, or larger, are preferred. Some applications, using rare or expensive analytes, may require volumes to be kept to a minimum. Fluidic cells with volumes as low as ~50 μL can be constructed. Microfluidic circuitry will be necessary if smaller volumes (~nL range) are needed. Microfluidic cells, however, require fluid flow and dead volumes would need to be considered in their design. The reduction in size of the fluidic cell also increases the electrolyte access resistance of the system ($R_{electrolyte}$), which needs to be carefully considered. In addition to limiting the bandwidth of the nanopore sensor (see Section 3.2.1), the electrolyte access resistance in series with the nanopore resistance acts as a voltage divider, which can result in a significant potential drop across the fluid, consequently introducing a systematic error on the applied voltage across the pore. Nevertheless, since R_{pore} is typically >10 MΩ, and most fluidic cell designs have $R_{electrolyte} \sim 0.1-1$ kΩ, $R_{electrolyte}/R_{pore} \ll 1$, so that the effect is negligible; unless electrodes are positioned in a microfluidic channel, many centimeters away from the pore membrane.

Some nanopore-sensing applications require additional design considerations. In particular electrically gated nanopores and nanopores equipped with embedded nanoelectrodes [60,64−68] require electrical connection(s). Figure 3.4B shows an exploded view of an electrically gated nanopore fluidic cell assembly. The gated nanopore is mounted on a custom-machined PCB to make electrical contact via a wirebond to the gate electrode. PDMS, applied with a custom-made single bristle loop brush, is used to insulate the electrical contact, preventing exposure to the electrolyte.

Performing nanopore recordings in combination with optical microscopy, either for optical tweezers measurements or to perform simultaneous fluorescence and electrical detection, generates another set of integration challenges. Interested readers are directed to Refs. [69−72] for detailed discussions on these topics.

3.2.3 Ag/AgCl electrodes

A potential difference between a pair of Ag/AgCl electrodes drives electrochemical reactions which results in ion transport through the nanopore. Ag/AgCl is used to convert the baseline ionic current signal into a flow of electrons detected by the measurement electronics. Ag/AgCl electrodes, which are made from a silver wire coated with silver chloride, are typically used in chloride-based electrolytes, as they are considered nonpolarizable, reversible electrodes. When an electric potential is applied across a pair of Ag/AgCl electrodes the following reversible electrochemical reactions take place:

$$Ag + Cl^{-} \rightleftharpoons AgCl + e^{-} \tag{3.2}$$

At the anode (i.e., positive electrode), Cl^{-} ions react with the Ag to produce AgCl plus an electron (e^{-}) which flows through the wire to the measurement electronics and back to the cathode, where the reverse reaction takes places. The electric potential of these electrodes depends on the effective concentration of Cl^{-} ions near the electrode surface, which can be approximated by the Nernst equation. A potential difference between a pair of electrodes will therefore exist if they are immersed in solution of different salt concentrations. One disadvantage of Ag/AgCl electrodes is that the solid AgCl can be exhausted by the current flow and bare Ag wire consequently exposed. This can lead to unpredictable surface reactions, DC offsets and instability in the measured current. Ag/AgCl electrodes should therefore be treated with care, and periodically cleaned and rechlorinated as described below. When used properly Ag/AgCl electrodes have the advantage of being nonpolarizable, which means that a Faradaic current can flow for any change in electric potential (i.e., no capacitive current at the electrode/electrolyte interface). In contrast, other electrode types may need to overcome an overpotential before a particular electrochemical reaction is allowed to happen in order for the current carried by the ions in solution to be converted into electrons in the wire.

Properly maintained Ag/AgCl electrodes immersed in KCl solutions do not contribute a significant amount of noise on the ionic current signal. Low-frequency peaks in the power spectral density (PSD) of the ionic current signal can occasionally be observed in the $0.01-1$ kHz range as a result of the mechanical vibrations of the electrodes, and due to acoustic pickup. To minimize this excess noise, the instrument should be positioned on a vibration isolation table, in an acoustically insulated box, the headstage of the recording instrument should be securely fixed in place and the electrodes should be firmly clamped and kept as short as possible. To minimize electrode resistance, the surface area of Ag/AgCl in contact with solution should be made as large as practically possible, ensuring accurate, stable current measurements (Box 3.2).

3.2.4 Noise pickup

Interfacing the nanopore fluidic cell to a computer-controlled current amplifier requires special consideration to enable low-noise recording. Computers and connected equipment (e.g., oscilloscope) can generate considerable electrical noise arising from ground loops, transients from switching power supplies, and radiative electrical pickup. Proper shielding and placement of equipment and cables can therefore be crucial.

As a general precaution, when assembling a nanopore recording setup, one should make equipment connections in sequence and verify after each connection that the noise level remains low. In some circumstances, the computer power supply should be replaced with a higher-end model employing superior quality components. To minimize ground loops, one can also try plugging the computer into different power sockets. This is also true of oscilloscopes or any other auxiliary equipment. In addition, to minimize electrical pickup, the nanopore fluidic cell and headstage should be well shielded, preferably in a grounded copper enclosure (Faraday cage). Commercial current amplifiers generally provide a "signal ground" which is isolated from the chassis and power ground for this purpose. It is recommended that two layers of shielding be used, as depicted in Figure 3.1. The holder tightly embedding the fluidic cell serves as the primary Faraday cage. A larger secondary Faraday cage should be used to shield the amplifier headstage connected to the electrodes immersed in the fluidic cell reservoirs. To avoid

---

BOX 3.2 PREPARING AND CARING FOR Ag/AgCl ELECTRODES

Standard material on techniques in electrophysiology provide guidelines for preparing or conditioning Ag/AgCl electrodes. Some basics steps are reviewed here:

Preparing

1. Cut a $\sim$ 5 cm-long $\sim$ 0.5 mm-diameter Ag wire. Solder a 1 mm gold-plated contact pin at one end. Encapsulate the Ag wire with PTFE heat shrinkable tubing, while leaving a few millimeter of bare Ag exposed.
2. Clean the exposed Ag wire by immersing in concentrated nitric acid for 30 s, rinse well with deionized water.
3. Immerse Ag wire in concentrated Clorox bleach for $\sim$ 1 h. Place in the dark while plating occurs. The plated electrode should be a uniform light to purplish/salmon gray color. Store dry, in a dark place until use. AgCl is light sensitive and light exposure tends to darken the electrode. The color of the AgCl surface should not affect the performance of the electrode since the Ag/AgCl interface is not reached by light. Alternatively, sintered Ag/AgCl pellets can be used, while making sure all Ag is properly encapsulated by PTFE heat shrinkable tubing. Note that longer replating time (overnight), for both pellet and wire electrodes, can be necessary if experiments at voltages >1 V are performed.

Cleaning

1. Routinely, the electrodes should be cleaned by rinsing with 95% ethanol, and soaking in water for a few minutes. For a deeper clean, the electrodes can be soaked in 10% HCl for a few minutes before rinsing with deionized water.
2. If the wire electrode or pellet is particularly dirty or corroded, a new surface can be exposed by immersing in 10% nitric acid for a few minutes to roughen the surface and remove contaminant, followed by a copious rinse with deionized water.
3. To replate AgCl, step 3 above should be repeated.

Removing DC Voltage Offsets

1. Short out electrode wires using an alligator clip and immerse the electrodes in 1 M KCl for 2 h.
2. If this fails, replate electrodes following the protocol above, and repeat step 1. Alternatively, apply an AC current ($\sim$ mA) with a frequency of 50−400 Hz for 1 min while electrodes are immersed in 1 M KCl.

---

ground loops, a single connection to the current amplifier "signal ground" should be made.

When a syringe pump is used to flow solution past the nanopore, electrolyte-filled perfusion tubing entering the secondary Faraday cage may act as antennas and pick up radiated noise. Although shielding of the tubing is possible, it is generally easier to insert in series with the tubing a drip-feed reservoir to separate, with an air-gap, the electrical continuity of the perfusion solution [42]. An example of a schematic of such an experimental setup is drawn in the supplementary material of reference [73].

Some nanopore experiments may require the temperature to be controlled. This is particularly true in nanopore-based force spectroscopy experiments, where single-molecule kinetics are especially temperature sensitive [74]. A thermoelectric (Peltier) controller may be used to efficiently regulate temperature. However, temperature regulation with thermoelectric elements may also introduce excess noise in the system as a result of the currents flowing through the thermoelectric module and the thermistors, as well as mechanical vibrations generated by fans. To avoid introducing noise, the Peltier element and thermistors must be placed outside the Faraday cages. If rapid temperature adjustments are not required, heating and cooling may be achieved by circulating temperature-controlled water to the holder of the fluidic cell with the temperature offset precalibrated. A schematic of an active temperature control system can be found in [75].

Command signals generated by digital-to-analog converters can also contribute significantly to the noise. These signals should be heavily filtered, when possible, to minimize noise. Using low gain settings on the command voltage of a commercial current amplifier can help reduce the noise pickup. A more in-depth discussion of extraneous electrical interferences, and how to minimize them, can be found in Ref. [42].

3.3 Low-current measurement techniques

Low-level current recordings requires specialized instrumentation and good measurement practices [42,76]. While standard patch-clamp amplifiers used in electrophysiology, such as the Axopatch 200B (Molecular Devices), are capable of ultra low-noise current measurements, it is important to understand how such measurements are realized and to identify the various sources of noise, in order to control and minimize their contribution, thereby ensuring the highest SNR in nanopore recording of single-molecule events.

3.3.1 Shunt resistor

The simplest way to measure currents is to measure the voltage drop across a resistor. This so-called *shunt resistance* technique to measure current is the operating principle of most digital multimeters (DMM). As illustrated in Figure 3.5, a power supply is used to set a potential difference, $V_{applied}$, between electrodes positioned on either side of a nanopore. The current flowing through the pore, I_{pore}, is measured from the voltage drop, $I_{pore}R_{shunt}$, across the shunt resistor, R_{shunt}.

The main issue with such design is that the potential across the membrane is reduced compare to $V_{applied}$ as a result of the voltage drop across R_{shunt}. This results in an error in the applied voltage that depends on the current flowing through the pore. To minimize this error, R_{shunt} should be kept as small as possible, thereby minimizing the voltage drop across it. However, this is incompatible with small current measurements, as Ohm's law requires the use of a large R for highest sensitivity. Consequently, maximum measurement accuracy requires minimizing this voltage drop, also referred to as the *voltage burden*, as it impacts low-level current measurements. Resistive feedback (discussed below), offers a simple solution to this problem.

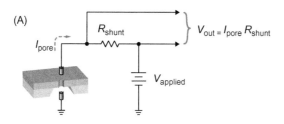

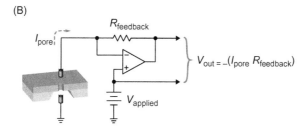

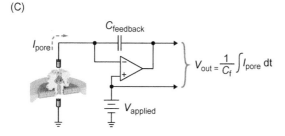

FIGURE 3.5

Current measuring circuits. (A) Shunt resistance technique. The current is measured as a potential drop across the resistor, using Ohm's law. The problem with this approach is that the voltage across the pore is not exactly equal to the applied potential, due to the potential drop across the shunt resistor. This voltage error increases as the shunt resistor is made large for high-sensitivity measurements. (B) Op-amp with a resistive feedback circuit. For all practical purposes the applied voltage at the inverting input terminal can be assumed to be equal to the applied voltage across the pore. (C) Op-amp with capacitive feedback circuit to measure very small currents (<100 pA). This scheme eliminates thermal noise from the feedback resistor and improves the dynamic range, though at the expense of frequent resets to discharge the capacitor, limiting the level of currents that can be measured. For this reason, it is more generally used with biological pores.

Source: Adapted from Ref. [77].

3.3.2 **Resistive feedback**

To circumvent the problem described above, an operational amplifier (op-amp) with a feedback resistor is usually used. Figure 3.5 shows a diagram of a simple model of a resistive feedback current-to-voltage converter circuit, also called a transimpedance amplifier. Such a circuit is typically found in most electrometers, picoammeters or electrophysiology amplifiers, including the Axopatch 200B (Molecular Devices), and the EPC-10 (HEKA). In an ideal op-amp the gain drives the output voltage until the voltage difference between the noninverting, V_+, and inverting, V_-, input terminals is zero. This implies that if V_+ is tied to a voltage source, then the other input, V_-, is very rapidly and precisely adjusted to the same potential. Following Figure 3.5, we can see that measurement of the output voltage of the op-amp is for practical purposes equivalent to measuring the voltage drop across the feedback resistor. This topology minimizes the voltage burden to typically a few hundred microvolts, while in comparison a DMM can have voltage burdens of tenths of volts. At the same time, an ideal op-amp has an infinite input resistance, i.e., it draws essentially no current through its input terminals. The current flowing out of the Ag/AgCl electrodes is thus equal to the current flowing through the feedback resistor. Following Ohm's law, we can therefore write the current flowing through the nanopore measured by the electrode as:

$$I_{pore} = -(V_{out} - V_{applied})/R_{feedback} \qquad (3.3)$$

In practice, commercial transimpedance amplifiers, such as the Axopatch 200B, make use of a subsequent "boost circuit" to remove the voltage offsets, and scale the gain (Box 3.3).

3.3.3 Capacitive feedback

In most cases, a resistive feedback topology is used to measure ionic current through solid-state nanopores, which is often in the range of $1-10$ nA for sub -10 nm pores under standard operating conditions (1 M KCl, $\sim$100 mV). However, when measuring smaller currents ($<$100 pA), as in the case of many biological pore applications, the thermal current noise of the feedback resistor can limit the SNR. In such a case, it may be better to measure the change in charge over time to determine the current. This is accomplished by replacing the feedback resistor with a capacitor, theoretically ensuring no thermal noise (see Section 3.4.1), and functions as an integrator. The capacitance value relates the voltage drop across the feedback capacitor to the measured charge, Q, as follows:

$$\frac{dQ}{dt} = C_{feedback} \frac{dV_{out}}{dt} \qquad (3.4)$$

For a constant applied voltage, the output voltage of the amplifier is related to the pore current by:

$$\frac{dV_{out}}{dt} = \frac{I_{pore}}{C_{feedback}} \qquad (3.5)$$

In addition to lowering the current noise, capacitive feedback allows for a faster settling time (only limited by the speed of the op-amp), as the speed of a resistive feedback circuit is limited to the $R_{feedback}C_{stray}$ time constant, which can be on the order of milliseconds or larger for high values of feedback resistances ($>1\ G\Omega$) or stray capacitance ($>1\ pF$), see Box 3.3.

The problem with capacitor-feedback topology is that large steady-state currents cannot be handled. Following Eq. (3.5), a constant current will cause the output voltage to ramp until saturation (typically ± 10 V). The feedback capacitor must therefore be periodically discharged to bring the output voltage back to zero. The duration of the discharge is typically $\sim 50\ \mu s$ for commercial amplifiers, and involves the use of switches, which can act as a source of noise. The frequency of discharge depends on the pore current and the value of the feedback capacitor. For a typical value of $C_{feedback} = 1$ pF, and a given steady-state current of 10 nA, $dV_{out}/dt = 0.01\ V/\mu s$, giving a period between discharge of 1 ms, and causing frequent sizeable voltage transients complicating the recording of such large currents.

BOX 3.3 DO IT YOURSELF I–V CONVERTER DESIGN RULES

Here we present important design rules for implementing the simple I–V converter depicted in Figure 3.5B, which can be applied for low-bandwidth current measurements (~ 1 kHz). The advantages of such measurement circuitry realized on a PCB, are its low-cost ($\sim\$100$), versatility (currents ranging from ~ 10 pA to ~ 1 mA can be monitored, and if needed voltages >1 V can be applied) and scalability (multiple current amplifiers can be integrated for gated nanopore or array measurements). The practical realization of such circuit requires a number of other error sources to be taken into account, which can affect the precision and accuracy of low-current measurements. These include:

- Leakage currents generated by stray resistance paths between the electrode wire connected to the measurement circuit and nearby points in the circuits at different potentials. Guarding is essential to reduce leakage currents, and is accomplished by surrounding the electrode wire with a guard at the same potential as the applied electrode voltage, as well as keeping the wire as short as possible.
- Input offset currents generated by the measurement circuit, when the input of the amplifier is left open. Careful choice of the op-amp connected as the I–V converter can minimize the offset current. The AD549 op-amp is a good choice, as the input bias current is on the order of ~ 100 fA.
- External offset currents generated by sources such as triboelectric and piezoelectric or stored charge effects. To minimize these effects, cables should be kept short and tied to a nonvibrating surface.
- A fundamental problem of a resistive feedback amplifier is that the output bandwidth is inherently low. The feedback resistance in parallel with the stray capacitance forms a one-pole RC filter with a -3 dB corner frequency approximated as $1/(2\pi R_{feedback}C_{stray})$. This filter limits the response time to typically milliseconds, and consequently distorts the frequency response of the current signal above a few kilohertz. To improve the frequency response of an I–V converter, one should therefore try to reduce C_{Stray}. Note that to overcome this limitation the Axopatch 200B and similar instruments make use of a high-frequency boost circuit with a frequency-dependent gain to correct the frequency response, which displaces the corner frequency to a much higher value (~ 100 kHz). For an example of a correcting circuit design see Ref. [77].

3.4 Bandwidth and background noise

Previous studies on bandwidth and background noise of ionic current recordings for patch clamping applications [38–40,42,78] as well as in submicron- and nano-pore systems have been published [25,26,55–57,61,73,79]. Here we provide a review of this material and examine the individual noise sources specifically in the context of nanopore recordings. In a properly configured nanopore recording instrument (see Section 3.2), the background noise sources typically originate from the measurement electronics, the nanopore device and through their integration. Figure 3.6A shows a simplified schematic of a nanopore device connected to a classical transimpedance circuit. When multiple noise sources are present in a complete recording system, the total root-mean-square (RMS) noise signal that results is the square root of the sum of the individual average mean-square noise values. This means that the worst noise source in the system will tend to dominate the total noise.

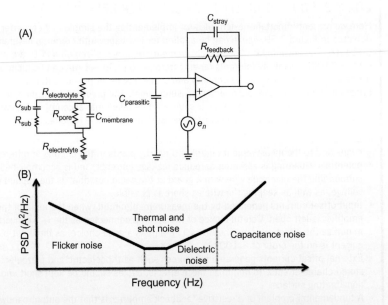

FIGURE 3.6

(A) Simplified circuit schematic of a nanopore device connected to transimpedance circuit. As described in Figure 3.2, a nanopore can be modeled as a resistance, R_{pore}, in series with an electrolytic access resistance, $R_{electrolyte}$, along with a capacitance from its freestanding membrane, $C_{membrane}$ and support structure, $C_{substrate}$. Additional parasitic capacitance, $C_{parasitic}$, is associated with the integration with the electronic circuitry, with contribution from wiring, the amplifier input, and the stray feedback capacitance, C_{stray}. e_n is the equivalent input voltage noise of the amplifier. (B) Illustration of the dominant sources of noise in the PSD of the ionic current as a function of frequency

Source: Adapted from Ref. [25].

As we will discuss in this section, analysis of bandwidth in nanopore record-ing reveals that the dominant sources of noise are associated with flicker noise in the low-frequency regime (typically <1 kHz) and dielectric loss and capaci-tance at higher frequencies. Figure 3.6B shows the PSD as a function of fre-quency, which illustrates these dominant sources of current noise. Different noise sources each have distinct frequency dependence. The total current noise PSD, S_I, can therefore be broken down into components and fitted to a polyno-mial of the form:

$$S_I(f) = \frac{a_0}{f} + a_1 + a_2 f + a_3 f^2 \text{ A}^2/\text{Hz} \tag{3.6}$$

where f is the frequency in Hz, and the terms a_0, a_1, a_2, a_3 represent the contribu-tions from the flicker (pink), thermal (white), dielectric (blue), and capacitance (purple) noises, respectively. We review each of these noise sources in detail below.

3.4.1 Low-frequency spectrum

The noise floor is set by thermal noise of the feedback resistor and/or the resistance of the nanopore, in the absence of a voltage applied across the membrane. However, following application of an electric potential, a low-frequency fluctuation, with $1/f$ characteristics (flicker noise), of the nanopore conductance appears as the dominant noise source, which impacts the stability of the ionic current recording.

3.4.1.1 Thermal noise

Thermal noise of the feedback resistor imposes a lower limit on the noise level of a transimpedance amplifier, while the thermal noise on the nanopore resistance sets the baseline noise of the nanopore recording instrument. This noise is gener-ated by the thermal fluctuations of charge carriers inside a conductive medium [80,81]. At equilibrium, the PSD of the thermal noise current is given by:

$$S_{\text{thermal}} = 4kT/R \tag{3.7}$$

where $4kT = 1.6 \times 10^{-20}$ Joule at room temperature, and R can either be the feed-back resistance or the resistance of the nanopore. The PSD is uniform throughout the frequency spectrum (i.e. white noise), and is independent of current. The ther-mal noise decreases as the value of the resistance increases as can be seen in Table 3.1. It is often useful to express the thermal current noise in terms of RMS noise (I_{thermal}) by taking the square root of the integral of the PSD. Since S_{thermal} is independent of frequency, I_{thermal} can be written as:

$$I_{\text{thermal}} = \sqrt{4kTB/R} \tag{3.8}$$

where B is the bandwidth in Hz, i.e., the frequency bounds of the integral. The RMS current noise increases as the square root of B and is inversely proportional to the square root of R. Thus, higher values of the feedback resistor and nanopores with larger resistance yield a lower thermal noise.

Note that, in practice, the feedback resistor of the transimpedance amplifier has appreciable stray capacitance as depicted in Figure 3.6A. If this stray capacitance is distributed evenly along the length of the resistor, the frequency response of the transimpedance amplifier will be affected, but will cause no change in the current noise. If the stray capacitance is distributed nonuniformly however, the noise can change dramatically, giving rise to frequency-dependent components.[1]

Table 3.1 Thermal, Shot, and Flicker Noise Amplitude Expressed as the PSD and the RMS Noise for a 10 and 100 kHz Bandwidth

	Resistance (MΩ)	$S_{thermal}$ (pA^2/Hz)	I_{RMS} (BW = 10 kHz) [pA]	I_{RMS} (BW = 100 kHz) [pA]
Thermal noise	10	1.66×10^{-3}	4.07	12.87
	100	1.66×10^{-4}	1.29	4.07
	1000	1.66×10^{-5}	0.41	1.29
	Current [nA]	S_{shot} [pA^2/Hz]	I_{RMS} (BW = 10 kHz) [pA]	I_{RMS} (BW = 100 kHz) [pA]
Shot noise	0.1	3.20×10^{-5}	0.57	1.79
	1	3.20×10^{-4}	1.79	5.66
	10	3.20×10^{-3}	5.66	17.90
	Current [nA]	S_{shot} at 1 Hz [pA^2/Hz]	I_{RMS} (BW = 10 kHz) [pA]	I_{RMS} (BW = 100 kHz) [pA]
Flicker noise	5	1 (low noise)	3.03	3.39
	5	100 (high noise)	30.35	33.93

Note that for white noise, the RMS noise is approximately one-sixth of the peak–peak noise. A normalized flicker noise amplitude of A = 4×10^{-8}, for a low 1/f-noise nanopore, and A = 4×10^{-6}, for high 1/f-noise nanopore, were used for the calculations (T = 300 K).

---

[1]More generally, the spectral density of the thermal current noise can be expressed in terms of the real part of the admittance of the device. It is simply $1/R$ for a resistor, but adopts a frequency-dependent form if the stray capacitance is nonuniformly distributed.

3.4.1.2 Shot noise

When current flows through a nanopore, another source of white noise is present, which is caused by random fluctuations in the motion of ions inside the channels. This so-called "shot noise" is present in any conducting medium and its PSD can be assumed to take the form:

$$S_{\text{shot}} = 2Iq \qquad (3.9)$$

where q is the effective charge of the current carrying species, and I is the average DC current. Typical estimated values for the shot noise in nanopore experiments are listed in Table 3.1.

3.4.1.3 Flicker noise

Applying an electric potential across the membrane gives rise to a low-frequency fluctuation (flicker noise or also known as pink noise) in the nanopore ionic current. In the low-frequency regime, this noise source largely dominates thermal and shot noise, especially in solid-state nanopores. Figure 3.7 depicts ionic current PSDs versus frequency calculated from ionic current recordings through a $\sim$9 nm solid-state nanopore exhibiting $1/f$-noise at $+200$ mV. This low-frequency conductance fluctuation with $1/f$ characteristic is a ubiquitous phenomenon found in a diversity of systems. It has been extensively investigated in nanopores, although no entirely satisfactory physical explanation has been developed [26,55,57,61,82–86]. The PSD is proportional to the square of the applied voltage across the membrane (or the square of the pore current) [82]. Smeets et al. [57,78] have shown that a model obeying Hooge's phenomenological relation [87], where noise scales inversely with the number of charge carriers present, best describes $1/f$-noise in solid-state nanopores. In such a case, the flicker noise PSD can be expressed as:

$$S_{\text{flicker}}(f) = \frac{AI^2}{f^\beta} \qquad (3.10)$$

where A is the normalized noise power defined as $A = \alpha/N_c$, in the Hooge model. N_c denotes the number of charge carriers, and α the Hooge parameter, which quantifies the level of flicker noise. The exponent β is typically unity, but has been observed to vary [26,55,61,84]. A constant value of $\alpha = 1.1 \times 10^{-4}$ was found for a wide range of salt concentrations ranging from $10^{-3}–10^0$ M.

It should be noted that there exists a large pore-to-pore variability in the level of the noise power (see Table 3.1), and although flicker noise cannot be completely removed, it can be minimized. This is illustrated in Figure 3.7, where a change of three orders of magnitude is observed. This implies that the above model may only describe flicker noise for low levels of $1/f$-noise ($S_I \sim 1$ pA^2/Hz at 1 Hz), and that other mechanisms leading to $1/f$ behavior may be simultaneously present [26]. These can include surface charge fluctuations, hydrophilic/hydrophobic effects, motion of charged constituents at pore walls, or anomalous

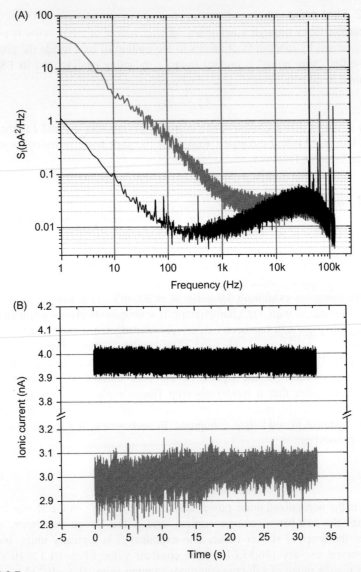

FIGURE 3.7

(A) Ionic current PSD versus frequency at +200 mV. The signal is low-pass filtered by a 4-pole Bessel filter at 100 kHz for a ∼9 nm pore in 1 M KCl buffered at pH7. The two curves show the difference in flicker noise amplitude between the partially wetted and the fully wetted case (i.e., before and after zapping). (B) Current recording at +200 mV of the same pore illustrating the difference in the current noise. Data in the current trace was decimated 10 times to reduce file size.

cooperative fluctuations of confined ions motion [82−85,88,89]. In practice, surface modifications involving atomic layer deposition [52,82] or piranha treatment [26] have been shown to improve $1/f$-noise levels. These results may be explained by an improvement in the surface hydrophilicity of the pore, removing potential nanobubbles [83] and facilitating wetting. This is supported by the observation that excessive flicker noise is found for pores with lower conductance values than expected (i.e., that are potentially only partially wetted) [57]. Indeed, recent work by Beamish et al. [62] demonstrated that the level of $1/f$-noise can be controlled by applying short pulses (∼100 ms) of moderately large potentials (∼10 V) across the pore. As shown in Figure 3.7, following this "*zapping*" method, a given nanopore, initially exhibiting low conductance and excessive level of flicker noise, can be seen to reduce its $1/f$-noise by nearly three orders of magnitude, while showing a slightly increased, though stable, conductance. Excessive flicker noise in solid-state nanopores can significantly complicate signal analysis and in a sizable fraction of pores render them useless for biological sensing applications, despite repeated cleaning procedures involving piranha or oxygen plasma treatments. The approach by Beamish et al. [62] may therefore provide an effective way to control $1/f$-noise levels in solid-state nanopores and improve their yield as single-molecule sensors.

3.4.1.4 *Protonation noise*

Following Bezrukov−Kasianowicz protonization noise in protein channels [90], Hoogerheide et al. [73] described another source of current noise in solid-state nanopores related to fluctuations of wall surface charges, which extends outside of the low-frequency regime. This excess noise is uniformly distributed and appears in the frequency range from 0.1 to 10 kHz. The PSD of this noise scales with the square of the current, and has been experimentally shown to vary with pH, and electrolyte concentration ($\propto c^{-3/2}$, for salt concentrations >50 mM). This noise can be minimized by operating nanopores at a pH value close to the point of zero charge (PZC), where the density of positive and negative surface functional groups are equal. For silicon nitride nanopores this value is between pH 4−6 [58,91,92].

3.4.2 **High-frequency spectrum**

Noise in this frequency regime generally sets an upper limit on the measurement bandwidth of single-molecule experiments in order to maintain an acceptable SNR. Dielectric properties of the nanopore chip's materials is the dominant noise source in this regime, up to frequencies of ∼10 kHz. As frequencies further increase, the dominant source of noise becomes the coupling of the amplifier's voltage noise with the total input capacitance, which, in most systems, is primarily determined by the design of the nanopore chip and its area exposed to liquids.

3.4.2.1 Dielectric noise

The dielectric noise is associated with the capacitance and the dielectric loss of the nanopore membrane and support structure. Nonideal dielectric materials dissipate some of the electrical energy as heat, resulting in the generation of additional thermal noise. The PSD of dielectric noise can be expressed as:

$$S_{\text{dielectric}}(f) = 8\pi kTDC_{\text{chip}} f \tag{3.11}$$

where $kT = 4.11 \times 10^{-21}$ J at room temperature, D is the dielectric loss, and C_{chip} the capacitance of the dielectric materials composing the membrane and support structure. Additional discussion regarding dielectric noise can be found in Refs [41,55,61].

3.4.2.2 Input capacitance noise

For frequencies >10 kHz the noise is dominated by the pairing of voltage noise at the input of the transimpedance amplifier and the sum of the capacitances distributed in the system. The PSD of the current caused by the input noise voltage is given by:

$$S_{\text{capacitance}}(f) = (2\pi f C_{\text{total}})^2 e_n^2 \tag{3.12}$$

where C_{total} is the total capacitance, and e_n is the equivalent voltage noise $[V/Hz^{1/2}]$ at the op-amp input. C_{total} is usually dominated by the distributed capacitance of the membrane and support chip, C_{chip}, though has contributions from the stray capacitance of the feedback resistor, C_{stray}, and other parasitic capacitances from the amplifier input and any connecting leads, $C_{\text{parasitic}}$, as depicted in Figure 3.6.

The value of the total input capacitance dramatically impacts the noise level when high-bandwidth (~100 kHz) measurements are required. In fact, solid-state nanopores fabricated on standard membranes, such as those commercially available as TEM windows, and mounted in standard fluidic cells have $C_{\text{chip}} \sim 0.5-1$ nF, and contribute a considerable amount of noise in the 10−100 kHz range, making single-molecule measurements at 100 kHz bandwidth often unfeasible. In the past few years, researchers have developed silicon nitride nanopore chips with reduced capacitance, significantly improving SNR in the high-frequency regime. These approaches include: painting PDMS [26] or photolithographically defining photoresist [55] on the nanopore chip to reduce the area exposed to solution; supporting the silicon nitride membrane with a thick silicon oxide layer to electrically isolate it from the doped silicon substrate [25,32,54−56]; and minimizing the area of the freestanding membrane (as discussed in Section 3.6).

3.5 Noise filtering, sampling, and resolution

Single-molecule studies using nanopores rely on the analysis of ionic current pulses. To obtain information on the structure, dynamics, and kinetics of biomolecules and biological processes under investigation, these electrical pulses must be distinguished from the background noise and statistically analyzed with respect to their depth (ΔI) and duration (Δt) [93,94]. A common way to increase SNR is to attenuate the high-frequency components of the ionic current signal. Reduction of the signal bandwidth is typically achieved by the use of active low-pass filters. Unfortunately, this unavoidably lowers the temporal resolution and distorts the shape of short pulses. This in turn can lead to a loss in structural information or other key biomolecular traits that would have otherwise been observable in the unfiltered data. The amplitude and phase of a signal are affected in different ways by the type of filters and the number of poles used (the larger the number of poles, the higher the steepness of the roll-off). Butterworth and Bessel filters are the two most commonly used filters. The Bessel filter should be used when studying current signals in the time domain, whereas Butterworth are preferred when analyzing signals in the frequency domain. This is because the former is well behaved in response to transients in the signal but does not provide as sharp a roll-off in the frequency domain (Figure 3.8).

Choosing the right cutoff frequency, f_c (defined as the frequency at which the signal is reduced by -3 dB, or $1/\sqrt{2}$), depends on the expected duration of the electrical pulses and the minimum acceptable SNR for an experiment. The rise-time, τ_{rise}, defined as the time taken for a signal to increase from 10% to 90% of its maximum value, can be expressed for a four-pole Bessel filter (as found in the Axopatch 200B) as:

$$\tau_{rise} \approx \frac{0.35}{f_c} \tag{3.13}$$

Because of the finite rise time of a filter, the electrical pulse edges of a single-molecule event will be distorted and the pulse amplitude will be attenuated if the pulse duration, $\tau_{event} \leq 2\tau_{rise}$. For $f_c = 100$ kHz, $\tau_{rise} = 3.5$ µs, such that events <7 µs cannot be accurately detected, with this particular low-pass cutoff frequency. In practice, a few more data points may be necessary to identify with confidence the plateau of the electrical pulse, without *a priori* knowledge of the pulse shape. Recently, Pedone et al. [95] introduced improved criteria to measure the width of short electrical pulses and proposed a method to recover the real pulse height, from the slope of the pulse's falling edge, which renders the identification of a pulse plateau unnecessary for events $\geq 0.2/f_c$. This method can significantly improve the accuracy of the data analysis beyond the usual limitation imposed by electronic filters. Note that many other signal-processing techniques exist, and can be used to accurately recover the pulse width and height of short electrical pulses [12,96], with various degree of complexity, including stochastic analysis using hidden Markov models [97,98], which are outside of the scope of this chapter.

A sufficient number of data points must be acquired to adequately sample the transient electrical pulses resulting from rapid single-molecule events. The choice

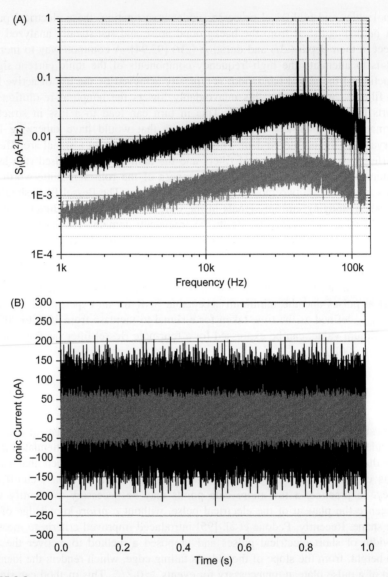

FIGURE 3.8

(A) Current PSD versus frequency at no applied voltage, low-pass filtered by a 4-pole Bessel filter at 100 kHz for two membranes with no nanopore immersed in 1 M KCl buffered at pH7. The two curves show the difference in high-frequency noise amplitude between a device structure with high (black, >500 pF) and low (gray, <100 pF) input capacitance. The "high" capacitance chip is a 30 nm-thick, 50 μm-wide TEM window, part# NT005X from Norcada mounted as shown in Figure 3.4A. The "low" capacitance chip is painted with a ~10 μm layer of PDMS up to ~50 μm around the freestanding membrane. (B) Current recording as a function of time, at no applied voltage for the same devices as in (A) illustrating the difference in the current noise.

of sampling frequency is influenced by the particular filter cutoff frequency selected. Any noise in the sampled signal with a frequency component greater than half the digitizing frequency, f_d, will appear in the digitized trace under the alias of lower-frequency noise. In other words, there is a "folding back" of these higher-frequency components into the frequency range from 0 to $f_\mathrm{d}/2$ [99]. The Nyquist principle states that this aliasing of the noise is introduced on a signal if the digitizing frequency is not at least twice the filter cutoff frequency:

$$f_\mathrm{d} \geq 2f_\mathrm{c} \tag{3.14}$$

The minimum sampling frequency is called the Nyquist frequency. In practice however, it is best to set the digitizing frequency significantly higher than the cutoff frequency (e.g., $f_d = 5f_c$ is common). Today's DAQ cards can achieve simultaneous acquisition of multiple signals at 500 kHz sampling frequency with 16-bit resolution, particularly field-programmable gate array (FPGA) based cards.

Analog-to-digital converters (digitizers) can also introduce quantization noise in a current trace. A 16-bit converter with a dynamic range set to ± 10 V, has a quantization step, δ, of $\delta = 20\ V/2^{16} = 305\ \mu V$. When the quantization is small relative to the signal being measured, the PSD of the quantizing voltage noise can be approximated by [42,84]:

$$S_\mathrm{digitization} = \frac{\delta^2}{6f_d} \tag{3.15}$$

According to the equation above, this excess quantization noise arising from digitization is uniformly distributed over the frequency spectrum. This noise can be safely ignored if it is negligible with respect to other sources of noise. One must therefore carefully adjust the dynamic range of the digitizer or use proper gain settings on the amplifier to ensure that that the signal being digitized is reasonably large relative to the quantization step.

3.6 Outlook: pushing the detection limit

Continuing progress of nanopore-based analysis of individual biomolecules demands that limits of detection be pushed by further improvements to bandwidth and SNR of nanopore recordings. As discussed in this chapter, today's nanopore setup can achieve temporal resolution of tens of microseconds. Nevertheless, at this maximum bandwidth, and depending on the application, most setups do not provide a minimum acceptable SNR >5 to accurately detect biomolecules. Improving the performance of nanopore sensors is particularly important for biomedically relevant applications focused on analysis of proteins and short DNA/RNA fragments [12,27−35] or differentiating individual nucleobases in genomic DNA [4,36−37]. In addition, more basic research endeavors would benefit from lower noise, higher bandwidth recordings, including the ability to study inside the nanoconfined geometry of a nanopore: surface dynamics; local chemical reactions; transport phenomena; or kinetics of protein structural transitions. While it is possible to hack a commercial current amplifier to boost its performance [39], as discussed in this chapter, the noise limitations of conventional nanopore recordings stem, in part, from the dielectric properties of the materials composing the nanopore support chip and its associated capacitance as well as the parasitic capacitances arising from the integration with the measurement electronics. Many research groups are actively tackling these engineering challenges.

3.6.1.1 *Solid-state nanopore devices with reduced capacitance*

Over the past few years, specifically designed nanopore chips have been fabricated in an effort to reduce capacitance compared to basic TEM window-type structures. Reducing the effective chip capacitance to improve the noise performance of solid-state devices was first explicitly revealed in [26]. Although the use of such improved nanopore chips is rapidly intensifying in nanopore-based sensing experiments [63,65], few articles actually describe the dielectric properties and capacitance of their nanopore support structure. In the supplementary information of their article [54], Gershow and Golovchenko mention achieving a total effective capacitance of 13 pF by inserting a $\sim$2 µm-thick silicon dioxide layer in between a 20 nm silicon nitride membrane and the silicon substrate, as well as reducing the size of the freestanding membrane to 20×20 µm^2 and of the area exposed to liquids by assembling the nanopore chip with microfabricated PDMS gaskets. A more in-depth investigation of the effect of the support structure architecture was performance Dimitrov et al. [55], who revealed an effective chip capacitance of $\sim$10 pF can be achieved by spin coating a $\sim$4 µm-layer of polyimide with a $\sim$10 µm-opening defined by photolithography, on top a 30 nm silicon nitride membrane. The authors also developed an external capacitance compensation circuit in order to improve the response time of the current signal to a change in the applied potential. Although the frequency response was significantly enhanced, considerable noise was introduced in the measurement, indicating that mitigation of parasitic capacitance through optimized chip design and miniaturization is the preferred path to low-noise, high-bandwidth recordings. Wanunu et al. [32] improved the dielectric properties of their nanopore chips by depositing a 5 µm-thick thermal silicon dioxide layer underneath a $\sim$40 nm silicon nitride membrane. In addition to lowering electrical noise, their nanopore chips were also designed to enhance the blockage amplitude originating from single-molecule events by using e-beam lithography to locally thin the membrane to sub-10 nm in a 250×250-nm^2 area, thus maintaining the overall membrane mechanical strength. Waggoner et al. [56] investigated the effect of silicon nitride and silicon oxide layers on top of a 50 nm silicon nitride membrane, and more specifically the effect of the doping level of the silicon substrate. Interestingly, they have found that using a more lightly doped silicon wafer resulted in much smaller capacitances in the silicon depletion layers, and consequently significantly reduced the RC time constant of the devices. A 0.75 mm-thick silicon wafer with

5000-Ω cm resistivity (P-doped, $N_d \approx 8 \times 10^{11}$) produced comparable results to more complex structures fabricated with thick dielectric layers above the silicon nitride membrane.

3.6.1.2 Integrated nanopores

The maximum operating bandwidth of most commercial low-noise current amplifiers, traditionally used in electrophysiology measurements, is $\sim$100 kHz. Driven by the need to improve the response time of the nanopore detector and its measurement bandwidth particularly for DNA sequencing applications, some recent research efforts have been aimed at the development of integrated-CMOS transimpedance amplifiers dedicated to nanopore sensing [100−102]. To maintain an acceptable SNR at bandwidths in the range of 0.1−1 MHz, parasitic capacitances in the whole system must be reduced to the 1−10 pF-level or below. This entails lowering all input capacitances, including contributions from wiring and interconnects between the nanopore and the measurement electronics. Integrating the nanopore device on a CMOS die, or positioning it in close proximity, can help accomplish this. Note that, monolithic integration of a CMOS amplifier and a nanopore device is a serious challenge, due to process incompatibilities and limitations of postprocessing of CMOS chips. Tightly integrated "two-chip" systems are therefore, in the near-term, the most promising path to low-noise high-bandwidth measurements [103].

This is exactly what Rosenstein et al. [25,101] recently described. They fabricated using CMOS processing a low-noise current amplifier equipped with an Ag/AgCl microelectrode and positioned it in a fluidic cell directly below a solid-state nanopore device. The integrated instrument is capable of performing measurements at $\sim$1 MHz bandwidth, while maintaining an exquisitely low RMS current noise of $\sim$155 pA. This remarkable performance was achieved by reducing all sources of parasitic capacitances to $\sim$7.4 pF. With additional fabrication efforts

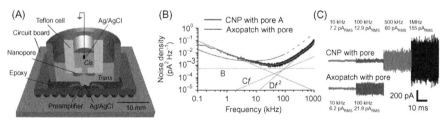

FIGURE 3.9

(A) Schematic of the integrated setup developed by Rosenstein et al., [25] dubbed CNP for CMOS-integrated nanopore platform. Measured ionic current PSD of a nanopore integrated with the CMOS amplifier at no applied bias. Membrane capacitance of pore A is 6 pF. (C) RMS current noise of the nanopore measured with the CMOS amplifier shown in (B).

Source: Adapted from Ref. [25]

the authors propose to further reduce the total input capacitance to ~1 pF, by reducing the contribution from the nanopore chip to below 0.5 pF. Figure 3.9 shows a schematic of the integrated nanopore platform developed, together with the measured current PSD versus frequency and current RMS at various measurement bandwidths.

The integrated nanopore-sensing platform fabricated by the Rosenstein model will undoubtedly promote further development of solid-state nanopore-based sensors interfaced with CMOS electronics. To continue to push the limit of detection, and reach a bandwidth >10 MHz, as required by some applications such as DNA sequencing, will be challenging. It will most certainly require improved amplifier topologies, entirely new nanopore chip designs, and drastically different integration approaches to keep parasitic capacitances to a minimum. Innovating solutions to these exciting engineering challenges will likely emerge following the development of new nanopore fabrication methods. Indeed, to fully leverage the semiconductor industry manufacturing capabilities to create arrays of low-noise high-bandwidth nanopore devices for high-throughput single-molecule electrical detection, alternative nanopore fabrication strategies will need to be developed to lower the complexity and cost associated with the use of beams of energetic particles [4]. These future nanopore fabrication methods will, in turn, guide the choice of materials and integration paths.

Ackowledgements

The author would like to thank Nahid Jetha for carefully reviewing the manuscript, and James Hedberg for the artwork. Many thanks to Eric Beamish, Kee Paik, Kyle Briggs, and Harold Kwok for their important contributions to the development of the protocols and data presented. This work was supported by the Natural Sciences and Engineering Research Council of Canada (NSERC).

References

[1] Nakane JJ, Akeson M, Marziali A. Nanopore sensors for nucleic acid analysis. J Phys Condens Matter 2003;15:R1365—93.

[2] Dekker C. Solid-state nanopores. Nat Nanotechnol 2007;2:209—15.

[3] Kasianowicz JJ, Robertson JWF, Chan ER, Reiner JE, Stanford VM. Nanoscopic porous sensors. Annu Rev Anal Chem (Palo Alto, Calif.) 2008;1:737—66.

[4] Branton D, Deamer DW, Marziali A, Bayley H, Benner SA, Butler T, et al. The potential and challenges of nanopore sequencing. Nat Biotechnol 2008;26:1146—53.

[5] Venkatesan BM, Bashir R. Nanopore sensors for nucleic acid analysis. Nat Nanotechnol 2011;6:615—24.

[6] Deamer DW, Akeson M. Nanopores and nucleic acids: prospects for ultrarapid sequencing. Trends Biotechnol 2000;18:147—51.

[7] Lemay SG. Nanopore-based biosensors: the interface between ionics and electronics. ACS Nano 2009;3:775−9.

[8] Aksimentiev A. Deciphering ionic current signatures of DNA transport through a nanopore. Nanoscale 2010;2:468−83.

[9] Kowalczyk SW, Tuijtel MW, Donkers SP, Dekker C. Unraveling single-stranded DNA in a solid-state nanopore. Nano Lett 2010;10:1414−20.

[10] Fologea D, Gershow M, Ledden B, McNabb DS, Golovchenko JA, Li J. Detecting single stranded DNA with a solid state nanopore. Nano Lett 2005;5:1905−9.

[11] Skinner GM, van den Hout M, Broekmans O, Dekker C, Dekker NH. Distinguishing single- and double-stranded nucleic acid molecules using solid-state nanopores. Nano Lett 2009;9:2953−60.

[12] Talaga DS, Li J. Single-molecule protein unfolding in solid state nanopores. J Am Chem Soc 2009;131:9287−97.

[13] Fologea D, Brandin E, Uplinger J, Branton D, Li J. DNA conformation and base number simultaneously determined in a nanopore. Electrophoresis 2007;28: 3186−92.

[14] van den Hout M, Skinner GM, Klijnhout S, Krudde V, Dekker NH. The passage of homopolymeric RNA through small solid-state nanopores 2011;7:2217−24.

[15] Tropini C, Marziali A. Multi-nanopore force spectroscopy for DNA analysis. Biophys J 2007;92:1632−7.

[16] Tabard-Cossa V, Wiggin M, Trivedi D, Jetha NN, Dwyer JR, Marziali A. Single-molecule bonds characterized by solid-state nanopore force spectroscopy. ACS Nano 2009;3:3009−14.

[17] Hornblower B, Coombs A, Whitaker RD, Kolomeisky A, Picone SJ, Meller A, et al. Single-molecule analysis of DNA-protein complexes using nanopores. Nat Methods 2007;4:315−7.

[18] Kowalczyk SW, Hall AR, Dekker C. Detection of local protein structures along DNA using solid-state nanopores. Nano Lett 2010;10:324−8.

[19] Bezrukov SM, Vodyanoy I, Parsegian VA. Counting polymers moving through a single ion channel. Nature 1994;370:279−81.

[20] Kasianowicz JJ, Brandin E, Branton D, Deamer DW. Characterization of individual polynucleotide molecules using a membrane channel. Proc Natl Acad Sci USA 1996;93:13770−3.

[21] Hall AR, Scott A, Rotem D, Mehta KK, Bayley H, Dekker C. Hybrid pore formation by directed insertion of α-haemolysin into solid-state nanopores. Nat Nanotechnol 2010;5:874−7.

[22] Maglia G, Restrepo MR, Mikhailova E, Bayley H. Enhanced translocation of single DNA molecules through alpha-hemolysin nanopores by manipulation of internal charge. Proc Natl Acad Sci USA 2008;105:19720−5.

[23] Stoddart D, Maglia G, Mikhalova E, Heron AJ, Bayley H. Multiple base-recognition sites in a biological nanopore: two heads are better than one. Angew Chem Int Ed Engl 2010;49:556−9.

[24] Craighead H. Future lab-on-a-chip technologies for interrogating individual molecules. Nature 2006;442:387−93.

[25] Rosenstein JK, Wanunu M, Merchant CA, Drndic M, Shepard KL. Integrated nanopore sensing platform with sub-microsecond temporal resolution. Nat Methods 2012;9:487−92.

[26] Tabard-Cossa V, Trivedi D, Wiggin M, Jetha NN, Marziali A. Noise analysis and reduction in solid-state nanopores. Nanotechnology 2007;18:305505.

[27] Storm AJ, Storm C, Chen J, Zandbergen H, Joanny JF, Dekker C. Fast DNA translocation through a solid-state nanopore. Nano Lett 2005;5:1193−7.

[28] Nanopore S, Storm AJ, Storm C, Chen J, Zandbergen H. Fast DNA translocation through a solid-state nanopore. 2005;1−5.

[29] Yusko EC, Pragkio P, Sept D, Rollings RC, Li J, Mayer M. Single particle characterization of Aβ oligomers in solution. ACS Nano 2012;6(7):5909−19.

[30] Yusko EC, Johnson JM, Majd S, Pragkio P, Rollings RC, Li J, et al. Controlling protein translocation through nanopores with bio-inspired fluid walls. Nat Nanotechnol 2011;6:253−60.

[31] Wei R, Gatterdam V, Wieneke R, Tampé R, Rant U. Stochastic sensing of proteins with receptor-modified solid-state nanopores. Nat Nanotechnol 2012;7:257−63.

[32] Wanunu M, Dadosh T, Ray V, Jin J, McReynolds L. Rapid electronic detection of probe-specific micro RNAs using thin nanopore sensors. Nat Nanotechnol 2010;5(11):807−14.

[33] Cressiot B, Oukhaled A, Patriarche G, Pastoriza-Gallego M, Betton JM, Auvray L, et al. Protein transport through a narrow solid-state nanopore at high voltage: experiments and theory. ACS Nano 2012;6(7):6236−43. doi:10.1021/nn301672g.

[34] Fologea D, Ledden B, McNabb DS, Li J. Electrical characterization of protein molecules by a solid-state nanopore. Appl Phys Lett 2007;91:539011−3.

[35] Wanunu M, Sutin J, Meller A. DNA profiling using solid-state nanopores: detection of DNA-binding molecules. Nano Lett 2009;9:3498−502.

[36] Cherf GM, Lieberman KR, Rashid H, Lam CE, Karplus K, Akeson M. Automated forward and reverse ratcheting of DNA in a nanopore at 5-Å precision. Nat Biotechnol 2012;30:344−8.

[37] Manrao EA, Derrington IM, Laszlo AH, Langford KW, Hopper MK, Gillgren N, et al. Reading DNA at single-nucleotide resolution with a mutant MspA nanopore and phi29 DNA polymerase. Nat Biotechnol 2012;30:349−53.

[38] Sakmann B, Neher E. pp. 700 Single-channel recording. Springer; 2009. Available from <http://www.springer.com/life+sciences/animal+sciences/book/978-0-306-44870-6>.

[39] Shapovalov G, Lester HA. Gating transitions in bacterial ion channels measured at 3 microns resolution. J Gen Physiol 2004;124:151−61.

[40] Hamill OP, Marty A, Neher E, Sakmann B, Sigworth FJ. Improved patch-clamp techniques for high-resolution current recording from cells and cell-free membrane patches. Pflugers Arch Eur J Phys 1981;391:85−100.

[41] Levis RA, Rae JL. The use of quartz patch pipettes for low noise single channel recording. Biophys J 1993;65:1666−77.

[42] The Axon guide—electrophysiology and biophysics laboratory techniques. Molecular Devices. The Axon Guide for Electrophysiology and Biophysics Laboratory Techniques, Third edition 2012. Available from <http://mdc.custhelp.com/euf/assets/content/Axon%20Guide%203rd%20edition.pdf>.

[43] Gouaux JE, Braha O, Hobaugh MR, Song L, Cheley S, Shustak C, et al. Subunit stoichiometry of staphylococcal α-hemolysin in crystals and on membranes: a heptameric transmembrane pore. Proc Natl Acad Sci USA 1994;91:12828 31.

[44] Wu M-Y, Krapf D, Zandbergen M, Zandbergen H, Batson PE. Formation of nanopores in a SiN/SiO$_2$ membrane with an electron beam. Appl Phys Lett 2005;87: 113106.

[45] Storm J, Chen JH, Ling XS, Zandbergen HW, Dekker C. Fabrication of solid-state nanopores with single-nanometre precision. Nat Mater 2003;2:537−40.

[46] Schneider GF, Kowalczyk SW, Calado VE, Pandraud G, Zandbergen HW, Vandersypen MK, et al. DNA translocation through graphene nanopores. Nano Lett 2010;10:3163−7.

[47] Garaj S, Hubbard W, Reina A, Kong J, Branton D, Golovchenko JA. Graphene as a subnanometre trans-electrode membrane. Nature 2010;467:190−3.

[48] Merchant CA, Healy K, Wanunu M, Ray V, Peterman N, Bartel J, et al. DNA translocation through graphene nanopores. Nano Lett 2010;10:2915−21.

[49] Venkatesan BM, Estrada D, Banerjee S, Jin X, Dorgan VE, Bae MH, et al. Stacked graphene-Al$_2$O$_3$ nanopore sensors for sensitive detection of DNA and DNA-protein complexes. ACS Nano 2012;6:441−50.

[50] Ikoma Y, Yahaya H, Kuriyama K, Sakita H, Nishino Y, Motooka T. Semiconductor nanopores formed by chemical vapor deposition of heteroepitaxial SiC films on SOI (100) substrates. J Vac Sci Technol B Microelectron Nanometer Struct 2011;29: 062001.

[51] Venkatesan BM, Shah AB, Zuo J-M, Bashir R. DNA sensing using nanocrystalline surface-enhanced Al$_2$O$_3$ nanopore sensors. Adv Funct Mater 2010;20:1266−75.

[52] Venkatesan BM, Dorvel B, Yemenicioglu S, Watkins N, Petrov I, Bashir R. Highly sensitive, mechanically stable nanopore sensors for DNA analysis. Adv Mater 2009;21:2771−6.

[53] Li J, Gershow M, Stein D, Brandin E, Golovchenko JA. DNA molecules and configurations in a solid-state nanopore microscope. Nat Mater 2003;2:611−5.

[54] Gershow M, Golovchenko JA. Recapturing and trapping single molecules with a solid-state nanopore. Nat Nanotechol 2007;2:775−9.

[55] Dimitrov V, Mirsaidov U, Wang D, Sorsch T, Mansfield W, Miner J, et al. Nanopores in solid-state membranes engineered for single molecule detection. Nanotechnology 2010;21:065502.

[56] Waggoner PS, Kuan AT, Polonsky S, Peng H, Rossnagel SM. Increasing the speed of solid-state nanopores. J Vac Sci Technol B Microelectro Nanometer Struct 2011;29:032206.

[57] Smeets RMM, Keyser UF, Dekker NH, Dekker C. Noise in solid-state nanopores. Proc Natl Acad Sci USA 2008;105:417−21.

[58] Smeets RM, Keyser UF, Krapf D, Wu MY, Dekker NH, Dekker C. Salt dependence of ion transport and DNA translocation through solid-state nanopores. Nano Lett 2006;6:89−95.

[59] Kowalczyk SW, Grosberg AY, Rabin Y, Dekker C. Modeling the conductance and DNA blockade of solid-state nanopores. Nanotechnology 2011;22:315101.

[60] Paik K-H, Liu Y, Tabard-Cossa V, Waugh MJ, Huber DE, Provine J, et al. Control of DNA capture by nanofluidic transistors. ACS Nano 2012;6:6767−75.

[61] Uram JD, Ke K, Mayer M. Noise and bandwidth of current recordings from submicrometer pores and nanopores. ACS Nano 2008;2:857−72.

[62] Beamish E, Kwok H, Tabard-Cossa V, Godin M. Precise control of the size and noise of solid-state nanopores using high electric fields. Nanotechnology 2012;23: 405301.

[63] Wanunu M, Meller A. Single-molecule analysis of nucleic acids and DNA-protein interactions. Single-molecule techniques: a laboratory manual. 2008. p. 395−420.

[64] Ivanov AP, Instuli E, McGilvery CM, Baldwin G, McComb DW, Albrecht T, et al. DNA tunneling detector embedded in a nanopore. Nano Lett 2011;11:279−85.

[65] Ayub M, Ivanov A, Instuli E, Cecchini M, Chansin G, McGilvery C, et al. Nanopore/electrode structures for single-molecule biosensing. Electrochim Acta 2010;55:8237−43.

[66] Jiang Z, Mihovilovic M, Chan J, Stein D. Fabrication of nanopores with embedded annular electrodes and transverse carbon nanotube electrodes. J Phys Condens Matter Inst Phys J 2010;22:454114.

[67] Nam S-W, Rooks MJ, Kim K-B, Rossnagel SM. Ionic field effect transistors with sub-10 nm multiple nanopores. Nano Lett 2009;9:2044−8.

[68] Jiang Z, Stein D. Charge regulation in nanopore ionic field-effect transistors. Phys Rev E 2011;83:.

[69] Soni GV, Singer A, Yu Z, Sun Y, McNally B, Meller A. Synchronous optical and electrical detection of biomolecules traversing through solid-state nanopores. Rev Sci Instrum 2010;81:014301.

[70] Keyser UF, van der Does J, Dekker C, Dekker NH. Optical tweezers for force measurements on DNA in nanopores. Rev Sci Instrum 2006;77:105105.

[71] Hong J, Lee Y, Chansin T, Edel JB, Demello AJ. Design of a solid-state nanopore-based platform for single-molecule spectroscopy. Nanotechnology 2008;19:165205.

[72] Chansin GA, Mulero R, Hong J, Kim MJ, DeMello AJ, Edel JB. Single-molecule spectroscopy using nanoporous membranes. Nano Lett 2007;7:2901−6.

[73] Hoogerheide DP, Garaj S, Golovchenko JA. Probing surface charge fluctuations with solid-state nanopores. Phys Rev Lett 2009;102:5−8.

[74] Jetha NN, Feehan C, Wiggin M, Tabard-Cossa V, Marziali A. Long dwell-time passage of DNA through nanometer-scale pores: kinetics and sequence dependence of motion. Biophys J 2011;100:2974−80.

[75] Jetha NN. Nanopore analysis of biological processes and macromolecules: low-velocity DNA translocation and prion protein conformational dynamics. Vancouver, Canada: University of British Columbia; 2013, Ph.D. thesis.

[76] Keithley JF. Low level measurements handbook, 6th edition 2008. Available from <http://www.keithley.com/knowledgecenter/knowledgecenter_pdf/LowLevMsHandbk_1.pdf>.

[77] Sigworth FJ. Electronic design of the patch clamp. Single-channel recording. Boston, MA: Springer; 2009. p. 95−127. [chapter 4].

[78] Levis RA, Rae JL. The use of quartz patch pipettes for low noise single channel recording. Biophys J 1993;65:1666−77.

[79] Smeets RMM, Dekker NH, Dekker C. Low-frequency noise in solid-state nanopores. Nanotechnology 2009;20:095501.

[80] Johnson J. Thermal agitation of electricity in conductors. Phys Rev 1928;32:97−109.

[81] Nyquist H. Thermal agitation of electric charge in conductors. Phys Rev 1928;32:110−3.

[82] Chen P, Mitsui T, Farmer DB, Golovchenko J, Gordon GR, Branton D. Atomic layer deposition to fine-tune the surface properties and diameters of fabricated nanopores. Nano Lett 2004;4:1333−7.

[83] Smeets R, Keyser U, Wu M, Dekker N, Dekker C. Nanobubbles in solid-state nanopores. Phys Rev Lett 2006;97:1−4.

[84] Siwy Z, Fuliński A. Origin of $1/f^{\alpha}$ noise in membrane channel currents. Phys Rev Lett 2002;89:1−4.

[85] Bezrukov SM, Winterhalter M. Examining noise sources at the single-molecule level: 1/f noise of an open maltoporin channel. Phys Rev Lett 2000;85:202−5.

[86] Danelon C, Santschi C, Brugger J, Vogel H. Fabrication and functionalization of nanochannels by electron-beam-induced silicon oxide deposition. Langmuir 2006;22: 10711−5.

[87] Hooge FN. 1/f noise is no surface effect. Phys Lett A 1969;29:139−40.

[88] Banerjee J, Verma MK, Manna S, Ghosh S. Self-organised criticality and 1/f noise in single-channel current of voltage-dependent anion channel. Europhys Lett (EPL) 2006;73:457−63.

[89] Tasserit C, Koutsioubas A, Lairez D, Zalczer G, Clochard M-C. Pink noise of ionic conductance through single artificial nanopores revisited. Phys Rev Lett 2010;105: 260602. Available from <http://prl.aps.org/abstract/PRL/v105/i26/e260602>.

[90] Bezrukov S, Kasianowicz J. Current noise reveals protonation kinetics and number of ionizable sites in an open protein ion channel. Phys Rev Lett 1993;70:2352−5.

[91] Butt H, Graf K, Kappl M. Physics and chemistry of interfaces. Wiley-VCH GmbH & Co; 2003. Available from <http://onlinelibrary.wiley.com/book/10.1002/3527602313>.

[92] Firnkes M, Pedone D, Knezevic J, Döblinger M, Rant U. Electrically facilitated translocations of proteins through silicon nitride nanopores: conjoint and competitive action of diffusion, electrophoresis, and electroosmosis. Nano Lett 2010;10:2162−7.

[93] Storm A, Chen J, Zandbergen H, Dekker C. Translocation of double-strand DNA through a silicon oxide nanopore. Phys Rev E, **71**. 2005.

[94] van den Hout M, Krudde V, Janssen XJA, Dekker NH. Distinguishable populations report on the interactions of single DNA molecules with solid-state nanopores. Biophys J 2010;99:3840−8.

[95] Pedone D, Firnkes M, Rant U. Data analysis of translocation events in nanopore experiments. Anal Chem 2009;81:9689−94.

[96] Colquhoun D, Sigworth FJ. Fitting and statistical analysis of single channel records. Single-channel recording. Boston, MA: Springer; 2009. p. 483−587. [chapter 19].

[97] Winters-Hilt S, Akeson M. Nanopore cheminformatics. DNA Cell Biol 2004;23: 675−83.

[98] Winters-Hilt S, Baribault C. A novel, fast, HMM-with-Duration implementation−for application with a new, pattern recognition informed, nanopore detector. BMC Bioinf 2007;8(Suppl. 7):S19.

[99] Heinemann SH. Guide to data acquisition and analysis. Single-channel recording. Boston, MA: Springer; 2009. p. 53−91. [chapter 3].

[100] Giehart BC, Howitt DG, Chen SJ, Zhu Z, Kotecki DE, Smith RL, et al. Nanopore with transverse nanoelectrodes for electrical characterization and sequencing of DNA. Sens Actuators B Chem 2008;132:593−600.

[101] Rosenstein J, Ray V, Drndic M, Shepard KL. Solid-state nanopores integrated with low-noise preamplifiers for high-bandwidth DNA analysis. 2011 IEEE/NIH Life Science Systems and Applications Workshop (LiSSA) 2011;59−62.

[102] Wang G, Dunbar WB. An integrated, low noise patch-clamp amplifier for biological nanopore applications. Conference proceedings: Annual International Conference of the IEEE Engineering in Medicine and Biology Society. IEEE Engineering in Medicine and Biology Society. Conference 2010, 2010;2718−21.

[103] Chen JWP, Howe RT. Wafer reconstitution with precision dry front-to-front registration. 15th Int. Conf. on Solid-State Sensors, Actuators and Microsystems.

Biological Pores on Lipid Bilayers

4

Joseph W.F. Robertson

Semiconductor and Dimensional Metrology Division,
National Institute of Standards and Technology, Gaithersburg, MD, USA

CHAPTER OUTLINE

4.1 Introduction

The field of nanopore sensing came into its own right with the development of biological ion channels as sensing elements [1,2]. These ion channels were developed as single-molecule-resistive pulse sensors by turning the black lipid membrane [3] into a platform to use the ion channel rather than as a platform to study ion channels.

The notion that ion channels function as nanoscopic Coulter counters [4,5] has been a mainstay in nanoporous sensor literature since early work by Kasianowicz and Bezrukov first demonstrated that ion channels can function as single-molecule sensors [4,6,7]. The concept caught on partially due to the simplicity of the Coulter counter. The core of Coulter's 1953 invention [8] is a concept that anyone who played with a garden hose as a child can easily understand. When a tube carrying a flowing fluid is pinched, squeezed, or otherwise constricted, the flow at the end of the hose is reduced. In a Coulter counter, the "fluid" is an electrolyte with ions current as the flow. When a particle enters the tube (Figure 4.1A), the ionic current through the pore is interrupted. With prior knowledge about the dimensions of the pore, the volume of an object can readily be estimated based on the change in conductance of the system through the relation:

$$\Delta R = \frac{4\rho}{\pi D} \left[\frac{\sin^{-1}(d/D)}{[1-(d/D)^2]^{1/2}} - \frac{d}{D} \right] \tag{4.1}$$

where ρ is the fluid resistivity, D is the diameter of the pore, d is the diameter of a *spherical* particle [9]. This technology opened a new world for the rapid and accurate sensing of cellular volume [9,10] and extension to the nanoscale [11] has enabled the rapid detection of colloids [12] and viruses [13] among other micro- and nanoscale particles. Although the analogy to the Coulter counter is an attractive one, biological nanopores based on nuances due to their miniscule size and complex chemistry go well beyond the realm of the Coulter counter (Figure 4.1).

It is possible to reduce the flow through the hose by occluding with a particle traversing the length of the pore (as in a Coulter counter). However, in a proteinaceous ion channel with length scales on the order of 5−10 nm, simple diffusion processes begin to limit the measurement capability. Einstein tells us that the time that it would take a particle to traverse the pore is on the order of 50−100 ns [14]. Currently, cutting-edge instrumentation can only achieve resolution in time on the order of 1 μs [15]. Thus, any particle (or molecule on this scale) that is to be detected by a nanopore must be retained within the pore through a direct physical interaction (e.g., adsorption [16], an entropic barrier [17,18], or another physical or chemical mechanism that can be designed or manipulated).

Continuing the garden hose analogy (Figure 4.1), ionic current can also be interrupted by applying force to the pore by applying an external force, thus constricting the channel from the outside. This feature can certainly be seen with mechanosensitive pores [19−21]. In addition to external forces, channels that serve as receptors (e.g., AB toxins [22]) can serve as rectifying resistive pulse sensors [23]. These receptor-based sensors can operate either by current interruptions induced by protein translocation or by a more mundane but equally effective capping of the ion channel.

While the Coulter-counter analogy is a useful first-order approximation protein nanopore sensors, it falls short in conveying the rich chemistry that is available— and necessary—for sensing in even the most simple pores. These interactions can

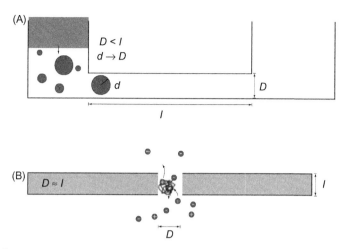

FIGURE 4.1

The subtle differences between Coulter counters and nanopore sensors are highlighted by the geometry. (A) In traditional Coulter counter, pressure-driven flow forces particles through a high aspect ratio pore. The extreme length of the pore requires particles to approach the diameter of the pore to detect meaningful resolution. (B) Protein nanopores have diameters that are in the same order of magnitude as the length such that volume filling provides a better determinant of the signal. In addition, because the size of the pore is commensurate with the size of the analytes, chemical interactions between the particle, pore, and electrolyte dictate the various observables including the magnitude of the current interruption and the residence time of the analyte in the pore.

be as simple as non-specific van der Waals interactions between a Gaussian chain polymer and hydrophilic residues within a nanopore [16,24−26], highly selective interactions between proteins that rely on the intricate geometric, and chemical alignment [23,27−30], or the intricate interactions between ions and a receptor site within a pore [7,31,32].

4.2 Formation: overview and experimental protocols

The fundamental processes that go into single protein electrophysiology have not changed in nearly 40 years [33,34]. Two reservoirs containing aqueous electrolyte solutions are separated by an insulating membrane in which a single protein nanopore or collection of nanopores short-circuits the insulator. An electrical potential is applied across the membrane by a battery (or power supply) connected to two ideal nonpolarizable electrodes, typically large, matched Ag/AgCl electrodes, and the current is measured with a high-impedance amplifier (Figure 4.2). Typical

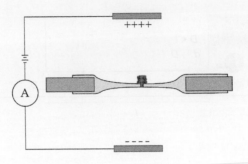

FIGURE 4.2

All biological nanopore measurement systems are comprised of a few simple components. An insulating membrane with a conductive nanopore separates two aqueous electrolyte solutions. Matched electrodes drives ionic current through connected to an external amplifier and power supply drives ionic current through the pore.

ionic currents are on the order of $10-100$ pA, which requires moderate shielding with a Faraday cage.

At the heart of the measurement is the black lipid membrane, which can be fabricated through a number of different methods. The most common technique is the solvent-free membrane self-assembly process developed by Montal and Mueller [3]. To form a membrane through this method, a hydrophobic partition (e.g., 25 μm thick Teflon) with a circular hole between 50 and 200 μm in diameter is treated with <0.5 μL of a 0.5% solution of hexadecane in pentane and blown dry with a stream of nitrogen. With the electrolyte solution level well below the membrane, lipid solution is added to each well such that ~5 monolayers of lipid are floating at the air−water interface. The solvent is allowed to evaporate and the lipid is allowed to equilibrate into a semiordered multilayer structure, ~10−15 min depending on the solvent. Then the solution levels are slowly raised above the hole in the partition spontaneously forming a bilayer. At this stage, small amounts of protein can be injected into solution (defined as the *cis* side) near the membrane and the current is monitored to detect pore formation. The amount of the protein must be calibrated to each individual preparation due to myriad unknowns including the degradation of the sample, concentration of active protein, and various other variables. When a single pore (or the appropriate number of pores for the measurement) is formed, the *cis* electrolyte is exchanged with fresh buffer. If this technique is performed properly, single-channel currents can be monitored for hours.

There are several variations on the method described above. To make membrane supports, White and colleagues [35−38] shrunk the aperture of the pore by fabricating lipid bilayers at the end of a glass capillary that was molded around a platinum scanning tunneling microscope tip. Using this method, channel containing membranes can be formed as small as 100 nm, which decreases the membrane-associated noise from both capacitance and mechanical fluctuations, while simultaneously improving the stability of the membrane. Another approach to membrane stabilization is to replace the biological (or biomimetic) lipid with a polymerizable component [39−41]. By first forming the bilayer, inserting the protein into the membrane and then polymerizing the interface measurements can be made on an individual protein for several weeks [40]. Yet another approach is to make membranes that can be formed quickly and efficiently through automated processes, for example, on devices fabricated through typical semiconductor processes, with the goal of making massively parallel measurements [42−44], or massively parallel droplet arrays [45,46].

4.3 Pore characterization: overview and experimental protocols

Biological nanopores are—without exception—transmembrane proteins. They are typically protein porins either made from transmembrane β-sheets—organized as a barrel, or α-helices that organize across a membrane with an aqueous cavity. As such, the tools that have been developed to study these nanopores have largely been developed through structural biology programs. Techniques of ubiquitous importance in structural biology include a diverse set of tools including electrophysiology, crystallography, high-resolution nuclear magnetic resonance (NMR) spectroscopy and electron paramagnetic resonance (EPR) spectroscopy, electron microscopy, atomic force microscopy (AFM), and neutron reflectometry/neutron diffraction [47].

4.3.1 Electrophysiological approaches

The impressive array of high-resolution techniques available can make the information available from "old" technology like classical electrophysiology seems trivial. However, proper experimental design and a bit of creativity can provide a wealth of information about critical dimensions of a nanopore and other relevant, critically important, details such as ion selectivity and chemistry.

4.3.1.1 Critical dimensions

These experiments start with the most simple assumptions that the water in the pore is disordered and bulk-like. In the absence of significant charge in the interior of the pore, Ohm's law can be used to estimate the radius of the pore. If we assume that the pore is a uniform cylinder with length (l) and radius (r), filled uniformly with an electrolyte solution of conductivity (σ), the conductance (g) of the pore should be:

$$g = \frac{\sigma \pi r^2}{l} \tag{4.2}$$

While a useful starting place, this approach is too simplistic to teach us much about the pore, and it fails miserably for pores that are so narrow that water begins to lose its bulk-like qualities (i.e., when the Debye length of the electrolyte begins to overlap with the walls of the nanopore).

For most pores, these simple assumptions do not hold in any meaningful way. For more realistic systems, pore geometry can be estimated by using finite-sized objects. The two most common approaches use either bulky ions such as tetraalkylammonium cations or nonelectrolyte polymers. In principle these techniques should be viewed as complimentary as the bulky ion approach tends to underestimate the pore diameter, and the polymer approach tends to overestimate the channel diameter. Ideally, these measurements will converge on the same result.

Both methods rely on the simple idea that ions or molecules that are small enough to fit inside the pore will affect the conductivity of the pore. For ions in the absence of any specific binding, ions that are small enough to partition into the pore will not substantially alter the observed current through the pore. Keeping the anion constant and testing the pore with cations of increasing size ($Me_4N^+ < Et_4N^+ < Pr_4N^+ < Bu_4N^+ < \cdots$), the current will drop by $\sim 2 \times$ when the cations are too large to partition into the pore. This technique was used early in the discovery and characterization of the PA63 channel to estimate a pore diameter > 1.1 nm [48].

The use of polymer probes to estimate the pore's geometry was developed by Krasilnikov in the early 1990s [49,50]. The principle is simple at first blush. The pore is sequentially tested with polymers of known hydrodynamic sizes. When the polymer is small enough to partition into the pore, the ionic current through the pore will decrease. By plotting the ratio of current (or conductance) with and without polymer present as a function of molecular weight, the polymer accessible radius of the pore can be estimated (Figure 4.3A). This limiting aperture should not be confused with an absolute measurement of the pore radius for a variety of reasons, as will be discussed in the following paragraphs.

To understand the limitations of polymer probes, it is important to understand the mechanism of how the current is blocked. As a first-order approximation, in bulk solution, the conductivity of electrolyte solution has a relatively straightforward dependence on the viscosity of the solution [52], and the viscosity of polymer solutions are strongly dependent on the concentration and size of the polymer molecule [53−55]. Thus, the presence of a nonelectrolyte polymer reduces the conductance of the electrolyte solution through a mechanism that is in proportion to the concentration of the polymer in solution. Because of the restrictive volume of a nanopore ~ 50 yL, a single-molecule partitioning into the pore will have a concentration on the order of 1 M—a substantial amount.

The digital approach described in Figure 4.3B works well for simple systems where little detail is needed, but this technique can be extended to understanding

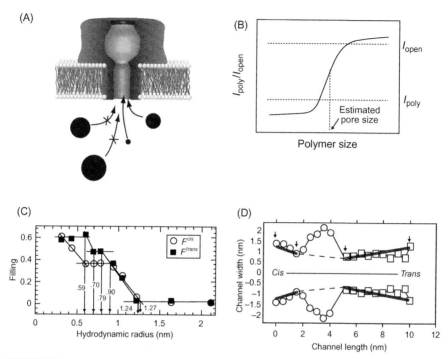

FIGURE 4.3

Nonelectrolyte polymers offer a course-grained approach to estimating nanopore topology. (A) An αHL channel spanning a lipid bilayer is tested with polymers with a known distribution of radius of gyration. (B) To a first approximation, large polymers are excluded from the pore, but small polymers partition into the pore reducing the conductance of the pore. (C) Monitoring the filling factor when polymer is added to each side of the pore allows the entry and narrowest constriction from either side of the pore to be estimated. (D) The observed polymer probe determined constriction for αHL (dashed line) is compared to the crystal structure (squares and circles).

Source: Parts (C) and (D) are reproduced from Ref. [51].

more detail of the pore geometry [56,57]. Instead of treating the data digitally (i.e., either in or out), a filling parameter (F) can be calculated:

$$F = \frac{[(g_0 - g_i)/g_i]}{[(\sigma_0 - \sigma_i)/\sigma_i]} \quad (4.3)$$

where g_0 is the single-channel conductance with polymer, g_i is the single-channel conductance in the absence of polymer, σ_0 is the conductivity electrolyte with polymer, and σ_i is the conductivity of electrolyte in the absence of polymer [56]. By adding the polymer to both sides of the membrane and the *cis* and *trans* sides separately it is possible to estimate some rudimentary structure within the channel (Figure 4.3C and D). Krasilnikov and colleagues [57] have applied this technique to several pores, perhaps most impressively, the results obtained through this method for α-hemolysin (αHL) faithfully reproduced the expected channel geometry based on the X-ray crystal structure [58] (Figure 4.3D).

Up to this point, the chemistry of the polymer probe does not significantly affect the results. For instance, the change in conductance in a Coulter counter is attributed exclusively to the volume of the particle with some geometrical constraints. Similarly, the coarse-grained polymer probe approach of Krasilnikov primarily uses empirical conductivity changes in bulk solution and extends this directly to the conductance in a nanopore [51]. The underlying physics is not critical for the efficacy of this technique. Regardless, a great deal can be learned by making simple observations with physical probes. However, with a little more work, we can greatly increase the resolution of the polymer probe techniques.

We can take a more refined approach by making a few small adjustments to the experiment and adding chemical detail to the polymer–pore system. Ordinarily, polymer synthesis results in a broad molecular weight distribution. Under most conditions, polymers and small molecules do not spend enough time in the pore to cause distinct current blockades [24]. However, if the system can be modified such that the residence time is extended from the nanosecond regime into the microsecond to millisecond regime, we can make measurements on individual molecules—negating the deleterious effects that are expected from a polydisperse sample. All mechanisms for increasing the residence time in the pore essentially reduce to increasing the activation barrier for the polymer to exit the pore.

There are many different ways to think of this problem, and are exquisitely laid out in books by DeGennes [59] and Muthukumar [60]. Essentially, the polymer needs to interact with the pore in a manner which allows it to be treated as a reversible reaction similar to that used by Woodhull to study acid–base equilibria in Na^+ channels [31]. The reaction between the polymer and the nanopore results in an energy barrier for the polymer extending the time necessary to make a low noise, high accuracy measurement.

Once we have established experimental conditions where long-lived isolated blockades can be observed (Figure 4.4A), we can develop a detailed picture of the physical and chemical processes that produce these results. On the scale of protein pores chemical interactions between the solvent, electrolyte and analyte greatly effect and can even dominate the response of the system [16,62,63]. When the conditions of the experiment are adjusted to optimize the polymer–pore interaction, such that the residence time of the polymer in the pore is $> \sim100\,\mu s$, the interaction can be characterized with a simple ratio of the blockade current $\langle i_0 \rangle$ divided by the open state current $\langle i \rangle$. A histogram of these values produces a spectrum of the polymer mixture where each peak is an individual n-mer (Figure 4.4C). With some knowledge of the polymer system under analysis,

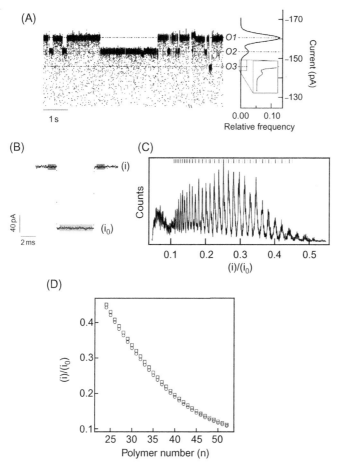

FIGURE 4.4

Single-molecule analysis can be used to refine pore size estimates with unprecedented precision. (A) Gating in the open state provides a baseline for the open channel conductance. (B) Analyzing the single-molecule blockade depths as the ratio of blocked to open current produces a spectrum where each peak is from a monomer-resolved polymer. (C) The shift in peak position from each open state can be used to provide exquisite resolution for geometrical changes in the pore geometry [61].

the nanopore can be modeled as a condensed phase single-molecule mass spectrometer (SMMS) [16,25]. For a polymer−nanopore system such as the classical polyethylene glycol (PEG) αHL system, the blockade depth is controlled by a three-part equilibrium reaction:

$$PEG + nK + \leftrightarrows (PEG - K) + n_{pore} \leftrightarrows (PEG - L) + n_{pore} \tag{4.4}$$

This reaction scheme has two major implications: a neutral polymer such as PEG can behave as if it were charged in solution and the polymer will mediate the flow of ions through adsorption–desorption reactions inside the pore. Thus, the dynamic equilibrium between the polymer, electrolyte, and pore allows for the determination of both size and charge of the analyte (i.e., a mass spectrometer). This negates the assumption that volume (or radius) is the controlling factor in a nanopore-occlusion experiment.

With a well-characterized polymer system, this single-molecule mass spectrometer can be turned around and used to characterize geometrical changes when a channel gates between two or more open states [61]. Using the ratio of the current in each of the open states:

$$R_{open} = \frac{\langle i_{high} \rangle}{\langle i_{low} \rangle} = \frac{1 + \delta L_{pore}}{1 + \delta A_{pore}} \tag{4.5}$$

where $\langle i_{high} \rangle$ and $\langle i_{low} \rangle$ are the mean current in the high current and low current open states, δL_{pore} and δA_{pore} are the change in length and area of the pore, respectively. The ratio of polymer-induced current blockades originating from each open state can then be used to provide a second equation with the geometrical changes:

$$R_{block,n} = \frac{r_{block,high,n}}{r_{block,low,n}} \approx 1 + \alpha(\delta L_{pore}) + \beta \left(\alpha + r_{block,high} \frac{L_{poly,n}}{L_{pore,high}} \right)(\delta A_{pore}) \tag{4.6}$$

where $r_{block,high,n}$ and $r_{block,low,n}$ are the peak positions of the open current normalized blockade depth for each n-mer, $\alpha = (1 - r_{block,high})$, and $\beta = ((A_{pore,high}/A_{poly,n})-1)^{-1}$. Using this methodology for the analysis of a gating αHL channel, the spectra of PEG ($n = 24-52$) was used to estimate changes in the β-barrel at the Å scale (i.e., $\delta L = 2.2$ Å, $\delta A = 1$ Å^2). The fine resolution afforded in this technique is unprecedented for detecting small changes in a functioning nanopore. However, it remains to be seen if these methods can be generalized to a wider number of channels and pores.

4.3.1.2 Selectivity

When ion channels can be isolated into an artificial membrane, determining the ionic selectivity is a relatively simple problem, which can be approached both experimentally [64] and theoretically [65,66]. From the experimental standpoint, the selectivity problem reduces to a simple application of the Nernst equation. Ionic current is measured with asymmetric ionic strength electrolytes on either side of the membrane typically two orders of magnitude (e.g., 0.01 M KCl cis|1 M KCl $trans$, or vice versa). The bias between electrodes is adjusted until the current is completely canceled (i.e., the reversal potential) and the Nernst equation then provides the relative permeability of each ion in the system. For porins typically used for single-molecule sensing, ion selectivity is typically low, but in some cases such as the PA63 channel, negative residues inside the channel produce a highly cation-selective pore [67].

4.3.2 **High-resolution structures**

The single biggest advantage of biological pores for sensing applications is the atomic precision afforded through conscripting biology for sensing applications. Additionally, structural biology has developed a number of techniques that can precisely determine the position and orientation of each residue within the pore (e.g., crystallography, NMR spectroscopy, and EPR spectroscopy). Although structures of membrane proteins are notoriously difficult to obtain, <400 have been solved to date [68,69]; the proteins typically used for nanopore sensing tend to be more amenable to crystallization than the average, and there are several high-resolution structures available for nanoporous proteins.

Despite the success in obtaining crystal structures for critical nanopore sensor proteins like *Staphylococcus aureus* αHL [58] and *Mycobacterium smegmatis* MspA [70] and *Escherichia coli* outer membrane porin *F* (OmpF) [71,72], there are several proteins important to sensing applications that are not amenable to crystallization such as *Bacillus anthraces* protective antigen [73], or that have unstructured regions that are not resolved in the crystal structure such as the gating loop of *Escherichia coli* OmpG [74,75] or OmpA [76].

The pinnacle for high-resolution structural data, x-ray crystallography, suffers from one particularly difficult limitation. The protein cannot be analyzed in a media that has direct relevance to biology or any sensing application. There are several alternative approaches to studying the structural elements of these protein channels. Nuclear and electronic spin techniques [77] currently offer the most compelling approaches as they both allow for biomimetic membrane systems and electrolyte systems during the measurement [78]. NMR spectroscopy in particular has struggled to become relevant for any protein system larger than ~30 kDa; although, recent advances in instrumentation, protein isotopic labeling [79–82], pulse sequences [83–91], and membrane/detergent systems [92–100] have extended the applicability of NMR to proteins as large as 1 MDa [91,101]. Despite the promise of NMR for determining structures in biomimetic interfaces for sensing applications, there are few NMR-derived structures of full-length porins. However, this is now a kinetic issue rather than a scientific issue at it is expected that many new structures will be solved in the coming years.

There are several other emerging techniques that do not provide high-resolution structures, but nevertheless deserve mention. Recent advances in AFM have enabled moderately high-resolution information of channel gating at video frame rates [102]. Using this and similar instrumentation it is possible to compare AFM images with crystal structures and infer the structure of gated states with only a crystal structure of the open channel [103]. AFM is particularly useful for protein systems that alter their structure that exists outside the membrane as changes on the nanometer scale can easily be resolved.

4.4 Bacterial pore-forming toxins

Although many different proteins have been investigated for their utility as single-molecule transducers, no species has shown more promise or success than the bacterial pore-forming toxins [1,2]. These proteins evolved in disease-causing bacteria as cell killers. Some, such as the hemolysin ion channels, particularly *Staphylococcus aureus* αHL, operate by forming relatively nonspecific pores in cell walls (e.g., red blood cells) inducing apoptosis through osmotic swelling [104–106]. Others of the AB toxin variety are composed of two components that can form channels that catalyze the translocation of the A component, an enzyme that catalyzes cell death through a variety of mechanism, and a B component that binds to receptors on a cell wall [22,73]. In many cases, the B component forms ion channel that can form the ultimate sensor for the presence of disease [23]. These systems, discussed below, offer the ultimate limit in both sensitivity and selectivity for a single-molecule biosensor. The selectivity, however, comes with a drawback that these pores tend to lack generalizability—a minor limitation in an era when membrane arrays are becoming the new state-of-the-art technique [43,44,46,107–109].

4.4.1 α-Hemolysin

αHL has a long history as the protein of choice for single-molecule sensing. Under high ionic strength conditions, the natural gating that occurs naturally in most ion channel experiments [110,111] shuts down [6,7]. Many different species of bacteria secrete pore-forming hemolysin pores including *Escherichia coli* [112] and *Staphylococcus aureus* [111,113]. The heptameric pore secreted by *Staphylococcus aureus* has become the workhorse protein for single-molecule sensing for a number of reasons. First and foremost, the pore forms a rigid, reproducible and perhaps most importantly a quiet ion channel [6,7,114]. While ideal electrical (ionic) properties of αHL make it a clear choice for sensor applications, the protein's stature is boosted by its known crystal structure [58] which is supported by electrophysiology experiments [111], with polymer probes [57], and with neutron reflectometry experiments at biomimetic interfaces [115].

The pore formed by αHL is a β-barrel protein formed by seven identical subunits of a 32 kDa protein. The protein is thought to form pores by diffusing to the membrane surface where heptameric prepore complexes assemble and then the β-barrel spontaneously partitions through the membrane forming a rigid cylindrical pore [116]. From the outside of the pore the protein best resembles a mushroom. The rigid pore (Figure 4.5) is often characterized as having three distinct sections. Beginning from the *cis* side of a membrane, defined as the size that protein was added to the solution, the pore has a "vestibule" which is a relatively large volume ~5 nm long with a biconic profile. At approximately the membrane surface, there is a narrow constriction that pinches down to ~1.5 nm in diameter separating the vestibule from the β-barrel, approximately a right regular cylinder

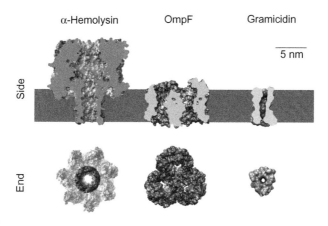

FIGURE 4.5

Protein nanopores have many different structural motifs to span a lipid bilayer membrane with an ionically conductive channel.

~5 nm long by 2.5 nm in diameter. The accessibility of the β-barrel is ideal for studying mid-sized polymers in the range of 1000 Da to >3500 Da where the polymer molecule can fully partition into the nanopore [16,24,25,51,117,118], as well as for both longer polymers and polymers that adopt less-coiled conformations such as nucleic acids or polyions [118−124]. Under some conditions, most likely due to pressure differentials from some membrane preparations, the β-barrel can step between several metastable states that are likely caused by the extension and narrowing of the pore similar to a Chinese finger trap [61].

αHL is stable under a wide range of conditions including temperatures below 0°C [125] to temperatures in excess of 90°C [126]. Even more remarkably, despite the presumed denaturation of the cap and vestibule under high concentrations of urea, the β-barrel retains its structure and remains useful for characterization of polymers such as large denatured proteins [124].

Unsurprisingly, the stability of the pore and its known crystal structure have made αHL an attractive target as a scaffold to engineer pores to either add binding sites for analytes [32,127−130], alter the ionic flow by changing the charge distribution [64] within the pore, or to add adapters into the pore to change the geometry and/or chemistry inside the pore [131−134].

4.4.2 Anthrax protective antigen

Protective antigen (PA63) from *Bacillus anthracis* is structurally homologous to αHL. The pore is formed by a heptameric [30,135], or occasional octameric [136], pore in the outer membrane of a cell when it initiates its infection process [137,138]. Through a process that is not well understood, PA63 catalyzes the translocation of two toxic proteins, lethal factor (LF) and edema factor (EF), across the cell wall, either of which results in cell death.

A cursory glance at the purported structure of PA63 compared to αHL suggests that the two pores could be complimentary to each other for applications that detect subtle differences in molecular mass (or volume/charge state of the analyte). For instance, αHL can easily detect differences in PEG at a resolution much better than the 44 g/mol of a single monomer. Perhaps, the narrower pore of PA63 channel could improve the resolution of the measurement and enable the technique to measure smaller polymers below the ~500 g/mol practical limit observed for αHL (unpublished observations). The answer is, somewhat surprisingly, no.

PA63 does not appear to behave like αHL for many of the analytes that have been examined. However, this does not make the channel useless as a sensor. Quite the contrary, as a central component of anthrax infection PA63 binds with a ~50 pM dissociation constant [23]. This is confirmed by the observation that intact lethal toxin and edema toxin (the complex of PA63 with LF and EF, respectively) can be found in the blood of infected animals [139]. This efficient binding offers two possible sensing strategies for the PA63 channel. The first is to use PA63 directly as a single-molecule sensor for anthrax infection [23]. Free heptameric PA63 pores produce a nearly Ohmic I−V curve, where LF- and EF-bound PA63 pores have a strongly rectifying appearance. Because the coupling of these proteins is so strong, the detector is essentially digital. In a similar vein, the PA63 channel can be used to screen for therapeutic agents against anthrax infection. The goal of such work is to design small molecule therapeutics that prevent the binding of LF and EF to PA63 [140,141].

4.5 Bacterial porins

4.5.1 Outer membrane porins

Omps are β-barrel proteins found in Gram-negative bacteria, mitochondria, and chloroplasts [142,143]. Unlike the pore-forming toxins described above, Omps are large β-barrels formed from a single peptide strand rather than intertwined multimers like αHL or PA63 (Figure 4.5). However, many Omps, like OmpG, can be found as monomers or dimers [74], or trimers like OmpF [144].

The natural function of these porins is to allow the transport of small molecules, such as nutrients or waste products in to or out of bacteria or mitochondria. Sensing applications began with the detection of single molecule−induced gating of OmpF, which is a known entry route for antibiotics [145]. Using a similar approach, Bezrukov and colleagues [146,147] have taken advantage of the lam B maltoporins selectivity for sugar transport into bacteria. The clearest application of such proteins is as sensitive and specific detectors for sugars. As with other membrane proteins, the selectivity of maltoporin is dictated by an exquisitely tailored binding site for maltose and maltodextrins [146,148]. Aside from serving as a conduit for sugars in and out of cells, maltoporin is also a receptor for lambda phage [149]. The lambda phage docking is particularly interesting as the trimeric pore detects phage binding with a stepwise decrease in current and the inhibition of sugar-translocation chattering. However, with maltose added to the side of the membrane opposite the virus, the pore can still detect the presence of the sugar even with virus bound.

4.5.2 *Mycobacterium smegmatis*: MspA

Recently, porins from mycobacteria have been studied as single-molecule sensors. Mycobacteria have a uniquely thick outer membrane comprised of mycolic acids—fatty acids with lengths up to 60 carbon atoms [150]. To span these membranes, these channels must have unusually large hydrophobic exteriors. Unique among these proteins is the MspA pore from *Mycobacterium smegmatis*. With their extremely thick outer membrane and porins such as MspA, *mycobacteria* have a high resistance to antibiotics [70,150,151].

The topology of MspA also makes it an ideal candidate for DNA sequencing applications [152,153]. MspA forms a large octameric pore that is best described as a goblet (Figure 4.6). MspA, as with other ideal single-molecule sensing pores, has a single central cavity where the ionic current and analyte molecules pass. The advantage of MspA is that the narrow constriction at the *trans* side of the channel is extremely short—1.2 nm in diameter and 0.6 nm long [152]. To achieve this effect with a protein, such as αHL, molecular adapters are necessary which add to the complexity of the system and increase the likelihood for failure (Figure 4.6).

4.6 Pore-forming peptides

Several different peptides have been shown to form pores in membranes. These peptides are produced by all organisms and often have antimicrobial properties through a variety of different mechanisms [154]. In part, because they are easy to prepare, purify, and modify, these pore-forming peptides have found use as nano-porous sensors.

4.6.1 Alamethicin

Unlike the rigid pore-forming proteins from the β-barrel channels above, alamethicin can form a dynamic range of pore sizes as that are attributed to channels made from aggregates of monomeric peptides [155], which can form multimeric pores [156]. Alamethicin pores have several conductance states which are attributed to pores incorporating increasing numbers of peptides [157]. The nature of this dynamic pore formation process requires fluid bilayers [41,155,158,159]. Polymer probes were used to estimate the lowest three conductance states at ~6, 9, and 11 nm [157]. Although the dynamic nature of these pores makes them

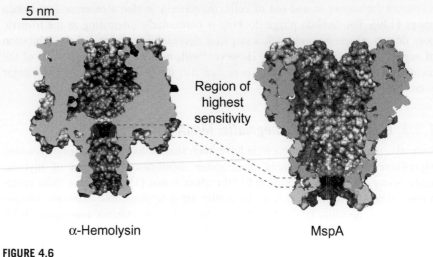

5 nm

Region of
highest
sensitivity

α-Hemolysin   MspA

FIGURE 4.6

A narrow constriction within a nanopore increases the resolution and sensitivity needed to
sequence DNA directly. αHL, modified with β-cyclodextrin and MspA both have
constrictions within their pore structure that focus the electric field creating an enhanced
sensing region. The constrictions in each protein are darkened (gray) to highlight their
location in the pore.

difficult for single-channel/single-molecule experiments, they were used as early
examples of "Nanopore Coulter Counters" [4].

4.6.2 Gramicidin

Gramicidin, unlike alamethicin, requires stable and fluid lipid bilayers for its
function. The gramicidin channel is a dimer that dynamically forms when two
peptides, each in a separate leaflet of the lipid bilayer, dimerize and form a pore
(Figure 4.5). This mechanism was confirmed by solid-state NMR of gramicidin-A
[160]. The transient nature of the gramicidin pore makes it a difficult protein for
single-channel sensing, but it has found application as an engineered sensor for a
number of different biological applications [161,162]. Because the peptide is
small, it can easily be modified with a number of different targets which alter the
channel conductance after an enzymatic reaction takes place at the surface [162].

4.7 Conclusion

Protein nanopores are currently the most versatile sensing elements for nanoporous sensors. These porins have evolved over billions of years for specific tasks,
which we can use directly (as with AB toxin sensors), or which we can hijack

for nonnatural uses (such as DNA sequencing and polymer spectroscopy). Part of the appeal of protein nanopores for sensing applications is the ability to rigorously control the chemistry within the pore to program the sensitivity and selectivity of the pore to individual analytes. Our knowledge of protein structure, with bioengineering, allows us to rigorously control most elements of the pore's chemistry including geometry and surface charge, and we can do this with atomic precision.

References

[1] Kasianowicz JJ, Robertson JWF, Chan ER, Reiner JE, Stanford VM. Nanoscopic porous sensors. Annu Rev Anal Chem 2008;1:737−66.

[2] Howorka S, Siwy Z. Nanopore analytics: sensing of single molecules. Chem Soc Rev 2009;38:2360−84.

[3] Montal M, Mueller P. Formation of bimolecular membranes from lipid monolayers and a study of their electrical properties. Proc Natl Acad Sci USA 1972;69:3561−6.

[4] Bezrukov SM, Vodyanoy I, Parsegian VA. Counting polymers moving through a single ion channel. Nature 1994;370:279−81.

[5] Bezrukov SM. Ion channels as molecular Coulter counters to probe metabolite transport. J Memb Biol 2000;174:1−13.

[6] Bezrukov S, Kasianowicz JJ. Current noise reveals protonation kinetics and number of ionizable sites in an open protein ion channel. Phys Rev Lett 1993;70:2352−5.

[7] Kasianowicz JJ, Bezrukov SM. Protonation dynamics of the alpha-toxin ion-channel from spectral analysis of pH-dependent current fluctuations. Biophys J 1995;69: 94−105.

[8] Coulter WH. A means for counting particles suspended in a fluid. U.S. patent number 2656508. 1953.

[9] Gregg EC, Steidley KD. Electrical counting and sizing of mammalian cells in suspension. Biophys J 1965;5:393.

[10] Rosenberg HM, Gregg EC. Kinetics of cell volume changes of murine lymphoma cells subjected to different agents *in vitro*. Biophys J 1969;9:592.

[11] DeBlois R, Bean C. Counting and sizing of submicron particles by resistive pulse technique. Rev Sci Instrum 1970;41:909.

[12] Saleh OA, Sohn LL. Quantitative sensing of nanoscale colloids using a microchip Coulter counter. Rev Sci Instrum 2001;72:4449−51.

[13] DeBlois R, Wesley R. Sizes and concentrations of several type-C oncornaviruses and bacteriophage-T2 by resistive-pulse technique. J Virol 1977;23:227−33.

[14] Einstein A. On the movement of small particles suspended in stationary liquids required by the molecular-kinetic theory of heat. Ann Phys 1905;17:549−60.

[15] Rosenstein JK, Wanunu M, Merchant CA, Drndic M, Shepard KL. Integrated nanopore sensing platform with sub-microsecond temporal resolution. Nat Meth 2012;9:487−94.

[16] Reiner JE, Kasianowicz JJ, Nablo BJ, Robertson JWF. Theory for polymer analysis using nanopore-based single-molecule mass spectrometry. Proc Natl Acad Sci USA 2010;107:12080−5.

[17] Berezhkovskii AM, Pustovoit MA, Bezrukov SM. Entropic effects in channel-facilitated transport: interparticle interactions break the flux symmetry. Phys Rev E 2009;80:020904.

[18] Muthukumar M. Polymer escape through a nanopore. J Chem Phys 2003;118: 5174—84.

[19] Perozo E, Cortes D, Sompornpisut P, Kloda A, Martinac B. Open channel structure of MscL and the gating mechanism of mechanosensitive channels. Nature 2002;418: 942—8.

[20] Betanzos M, Chiang C-S, Guy HR, Sukharev S. A large iris-like expansion of a mechanosensitive channel protein induced by membrane tension. Nat Struct Biol 2002;9:704—10.

[21] Martinac B. Bacterial mechanosensitive channels as a paradigm for mechanosensory transduction. Cell Physiol Biochem 2011;28:1051—60.

[22] Badizadegan K, Collier RJ, Lencer W. Membrane translocation by bacterial AB toxins. Method Microbiol 2002;31:277—96.

[23] Halverson KM, Panchal RG, Nguyen TL, Gussio R, Little SF, Misakian M, et al. Anthrax biosensor, protective antigen ion channel asymmetric blockade. J Biol Chem 2005;280:34056—62.

[24] Bezrukov SM, Vodyanoy I, Brutyan R, Kasianowicz JJ. Dynamics and free energy of polymers partitioning into a nanoscale pore. Macromolecules 1996;29: 8517—22.

[25] Robertson JWF, Rodrigues CG, Stanford VM, Rubinson KA, Krasilnikov OV, Kasianowicz JJ. Single-molecule mass spectrometry in solution using a solitary nanopore. Proc Natl Acad Sci USA 2007;104:8207—11.

[26] Bezrukov SM, Krasilnikov OV, Yuldasheva LN, Berezhkovskii AM, Rodrigues CG. Field-dependent effect of crown ether (18-Crown-6) on ionic conductance of α-hemolysin channels. Biophys J 2004;87:3162—71.

[27] De S, Olson R. Crystal structure of the *Vibrio cholerae* cytolysin heptamer reveals common features among disparate pore-forming toxins. Proc Natl Acad Sci USA 2011;108:7385—90.

[28] Zhang RG, Scott DL, Westbrook ML, Nance S, Spangler BD, Shipley GG, et al. The 3-dimensional crystal-structure of cholera-toxin. J Mol Biol 1995;251: 563—73.

[29] Merritt EA, Pronk SE, Sixma TK, Kalk KH, Vanzanten B, Hol W. Structure of partially activated *Escherichia coli* heat-labile enterotoxin (Lt) at 26-angstrom resolution. FEBS Lett 1994;337:88—92.

[30] Petosa C, Collier RJ, Klimpel K, Leppla S, Liddington R. Crystal structure of the anthrax toxin protective antigen. Nature 1997;385:833—8.

[31] Woodhull AM. Ionic blockage of sodium channels in nerve. J Gen Physiol 1973;61:687—708.

[32] Braha O, Walker B, Cheley S, Kasianowicz JJ, Song L, Gouaux JE, et al. Designed protein pores as components for biosensors. Chem Biol 1997;4:497—505.

[33] Neher E, Sakmann B. Single-channel currents recorded from membrane of denervated frog muscle-fibers. Nature 1976;260:799—802.

[34] Hille B. Ionic channels of excitable membranes. Sunderland, MA: Sinauer Associates Inc; 1992.

[35] White RJ, Zhang B, Daniel S, Tang JM, Ervin EN, Cremer PS, et al. Ionic conductivity of the aqueous layer separating a lipid bilayer membrane and a glass support. Langmuir 2006;22:10777—83.

[36] White RJ, Ervin EN, Yang T, Chen X, Daniel S, Cremer PS, et al. Single ion-channel recordings using glass nanopore membranes. J Am Chem Soc 2007;129: 11766−75.

[37] Zhang B, Galusha J, Shiozawa PG, Wang G, Bergren AJ, Jones RM, et al. Bench-top method for fabricating glass-sealed nanodisk electrodes, glass nanopore electrodes, and glass nanopore membranes of controlled size. Anal Chem 2007;79:4778−87.

[38] Schibel AEP, Edwards T, Kawano R, Lan W, White HS. Quartz nanopore membranes for suspended bilayer ion channel recordings. Anal Chem 2010;82: 7259−66.

[39] Shenoy DK, Barger WR, Singh A, Panchal RG, Misakian M, Stanford VM, et al. Functional reconstitution of protein ion channels into planar polymerizable phospholipid membranes. Nano Lett 2005;5:1181−5.

[40] Heitz BA, Xu J, Hall HK, Aspinwall CA, Saavedra SS. Enhanced long-term stability for single ion channel recordings using suspended poly(lipid) bilayers. J Am Chem Soc 2009;131:6662.

[41] Heitz BA, Jones IW, Hall HK, Aspinwall CA, Saavedra SS. Fractional polymerization of a suspended planar bilayer creates a fluid, highly stable membrane for ion channel recordings. J Am Chem Soc 2010;132:7086−93.

[42] Brueggemann A, Stoelzle S, George M, Behrends JC, Fertig N. Microchip technology for automated and parallel patch-clamp recording. Small 2006;2:840−6.

[43] Baaken G, Sondermann M, Schlemmer C, Ruehe J, Behrends JC. Planar microelectrode-cavity array for high-resolution and parallel electrical recording of membrane ionic currents. Lab Chip 2008;8:938−44.

[44] Baaken G, Ankri N, Schuler A-K, Rühe J, Behrends JC. Nanopore-based single-molecule mass spectrometry on a lipid membrane microarray. ACS Nano 2011;5: 8080−8.

[45] Malmstadt N, Nash MA, Purnell RF, Schmidt JJ. Automated formation of lipid-bilayer membranes in a microfluidic device. Nano Lett 2006;6:1961−5.

[46] Thapliyal T, Poulos JL, Schmidt JJ. Automated lipid bilayer and ion channel measurement platform. Biosens Bioelectron 2011;26:2651−4.

[47] Robertson JWF, Kasianowicz JJ, Banerjee S. Analytical approaches for studying transporters, channels, and porins. Chem Rev 2012;112:6227−49.

[48] Blaustein R, Finkelstein A. Diffusion limitation in the block by symmetrical tetra-alkylammonium ions of anthrax toxin channels in planar phospholipid bilayer membranes. J Gen Physiol 1990;96:943−57.

[49] Krasilnikov OV, Sabirov RZ, Ternovsky VI, Merzliak PG, Muratkhodjaev JN. A simple method for the determination of the pore radius of ion channels in planar lipid bilayer membranes. FEMS Microbiol Immun 1992;105:93−100.

[50] Krasilnikov OV. Sizing channels with neutral polymers. In: Kasianowicz JJ, Kellermayer MS, Deamer DW, editors. Structure and dynamics of confined polymers. Dordrecht, The Netherlands: Kluwer Academic Publishers; 2002.

[51] Krasilnikov OV, Bezrukov SM. Polymer partitioning from nonideal solutions into protein voids. Macromolecules 2004;37:2650−7.

[52] Boucher E, Hines PM. Properties of aqueous salt solution of poly(ethylene oxide)—thermodynamic quantities based on viscosity and other measurements. J Polym Sci Pol Phys 1978;16:501−11.

[53] Huggins M. The viscosity of dilute solutions of long-chain molecules IV dependence on concentration. J Am Chem Soc 1942;64:2716−8.

[54] DeBye P, Bueche A. Intrinsic viscosity, diffusion, and sedimentation rate of polymers in solution. J Chem Phys 1948;16:573–9.

[55] Mooney M. On an indeterminate integral in Einstein's theory of the viscosity of a suspension. J Appl Phys 1954;25:406–7.

[56] Krasilnikov OV, Da Cruz J, Yuldasheva L, Varanda W, Nogueira R. A novel approach to study the geometry of the water lumen of ion channels: colicin Ia channels in planar lipid bilayers. J Memb Biol 1998;161:83–92.

[57] Merzlyak PG, Yuldasheva LN, Rodrigues CG, Carneiro C, Krasilnikov OV, Bezrukov SM. Polymeric nonelectrolytes to probe pore geometry: application to the alpha-toxin transmembrane channel. Biophys J 1999;77:3023–33.

[58] Song L, Hobaugh M, Shustak C, Cheley S, Bayley H, Gouaux J. Structure of staphylococcal alpha-hemolysin, a heptameric transmembrane pore. Science 1996;274: 1859–66.

[59] de Gennes P-G. Scaling concepts in polymer physics. Ithaca, NY: Cornell University Press; 1979.

[60] Muthukumar M. Polymer translocation. Boca Raton, FL: CRC Press; 2009.

[61] Robertson JWF, Kasianowicz JJ, Reiner JE. Changes in ion channel geometry resolved to sub-ångström precision via single molecule mass spectrometry. J Phys Condens Mat 2010;22:454108.

[62] Rodrigues CG, Machado DC, Chevtchenko SF, Krasilnikov OV. Mechanism of KCl enhancement in detection of nonionic polymers by nanopore sensors. Biophys J 2008;95:5186–92.

[63] Kasianowicz JJ, Reiner JE, Robertson JWF, Henrickson SE, Rodrigues C, Krasilnikov OV. Nanopore-based technology: single molecule characterization and DNA sequencing. In: Gracheva M, editor. Detecting and characterizing individual molecules with single nanopores. New York, NY: Springer Verlag; 2012. p. 3–20.

[64] Merzlyak P, Capistrano M, Valeva A, Kasianowicz JJ, Krasilnikov O. Conductance and ion selectivity of a mesoscopic protein nanopore probed with cysteine scanning mutagenesis. Biophys J 2005;89:3059–70.

[65] Allen TW, Andersen OS, Roux B. Molecular dynamics—potential of mean force calculations as a tool for understanding ion permeation and selectivity in narrow channels. Biophys Chem 2006;124:251–67.

[66] Corry B. Understanding ion channel selectivity and gating and their role in cellular signalling. Mol BioSyst 2006;2:527.

[67] Blaustein R, Koehler TM, Collier RJ, Finkelstein A. Anthrax toxin—channel forming activity of protective antigen in planar phospholipid bilayers. Proc Natl Acad Sci USA 1989;86:2209–13.

[68] White SH. Biophysical dissection of membrane proteins. Nature 2009;459:344–6.

[69] White SH, editor. Membrane proteins of known 3D structure. Paper by Stephen White Laboratory, UC Irvine <http://blanco.biomol.uci.edu/mpstruc/listAll/list>; 2013 [accessed 05.02.13].

[70] Faller M, Niederweis M, Schulz G. The structure of a mycobacterial outer-membrane channel. Science 2004;303:1189–92.

[71] Cowan SW, Schirmer T, Rummel G, Steiert M, Ghosh R, Pauptit RA, et al. Crystal structures explain functional properties of two E. coli porins. Nature 1992;358:727–33.

[72] Cowan SW, Garavito RM, Jansonius JN, Jenkins JA, Karlsson R, König N, et al. The structure of OmpF porin in a tetragonal crystal form. Struct Fold Design 1995;3:1041–50.

[73] Young JAT, Collier RJ. Anthrax toxin: receptor binding, internalization, pore formation, and translocation. Annu Rev Biochem 2007;76:243−65.

[74] Mari SA, Köster S, Bippes CA, Yildiz O, Kühlbrandt W, Muller DJ. pH-Induced conformational change of the beta-barrel-forming protein OmpG reconstituted into native *E. coli* lipids. J Mol Biol 2010;396:610−6.

[75] Liang B, Tamm LK. Structure of outer membrane protein G by solution NMR spectroscopy. Proc Natl Acad Sci USA 2007;104:16140−5.

[76] Arora A, Abildgaard F, Bushweller J, Tamm L. Structure of outer membrane protein a transmembrane domain by NMR spectroscopy. Nat Struct Biol 2001;8:334−8.

[77] Hellmich UA, Glaubitz C. NMR and EPR studies of membrane transporters. Biol Chem 2009;390:815−34.

[78] Opella S, Marassi F. Structure determination of membrane proteins by NMR spectroscopy. Chem Rev 2004;104:3587−606.

[79] Grzesiek S, Anglister J, Ren H, Bax A. C-13 line narrowing by H-2 decoupling in H-2/C-13/N-15-enriched proteins—application to triple-resonance 4d J-connectivity of sequential amides. J Am Chem Soc 1993;115:4369−70.

[80] Tugarinov V, Kay LE. An isotope labeling strategy for methyl TROSY spectroscopy. J Biomol NMR 2004;28:165−72.

[81] Tugarinov V, Kay LE. Methyl groups as probes of structure and dynamics in NMR studies of high-molecular-weight proteins. ChemBioChem 2005;6:1567−77.

[82] Tugarinov V, Kay LE. Ile, Leu, and Val methyl assignments of the 723-residue malate synthase G using a new labeling strategy and novel NMR methods. J Am Chem Soc 2003;125:13868−78.

[83] Tugarinov V, Kay LE. 1H,13C-1H,1H dipolar cross-correlated spin relaxation in methyl groups. J Biomol NMR 2004;29:369−76.

[84] Tugarinov V, Hwang PM, Ollerenshaw JE, Kay LE. Cross-correlated relaxation enhanced 1H − 13C NMR spectroscopy of methyl groups in very high molecular weight proteins and protein complexes. J Am Chem Soc 2003;125:10420−8.

[85] Tugarinov V, Sprangers R, Kay LE. Line narrowing in methyl-TROSY using zero-quantum 1H−13C NMR spectroscopy. J Am Chem Soc 2004;126:4921−5.

[86] Tugarinov V, Kay LE. Quantitative NMR studies of high molecular weight proteins: application to domain orientation and ligand binding in the 723 residue enzyme malate synthase G. J Mol Biol 2003;327:1121−33.

[87] Bax A, Davis DG. Practical aspects of two-dimensional transverse. NOE spectroscopy. J Magn Reson 1985;63:207−13.

[88] Riek R, Fiaux J, Bertelsen E, Horwich A, Wüthrich K. Solution NMR techniques for large molecular and supramolecular structures. J Am Chem Soc 2002;124:12144−53.

[89] Pervushin K, Riek R, Wider G, Wüthrich K. Attenuated T-2 relaxation by mutual cancellation of dipole−dipole coupling and chemical shift anisotropy indicates an avenue to NMR structures of very large biological macromolecules in solution. Proc Natl Acad Sci USA 1997;94:12366−71.

[90] Pervushin K, Wider G, Riek R, Wüthrich K. The 3D NOESY-[H-1,N-15,H-1]-ZQ-TROSY NMR experiment with diagonal peak suppression. Proc Natl Acad Sci USA 1999;96:9607−12.

[91] Tzakos AG, Grace CRR, Lukavsky PJ, Riek R. NMR techniques for very large proteins and RNAs in solution. Annu Rev Biophys Biomol Struct 2006;35:319−42.

[92] Vold RR, Prosser RS. Magnetically oriented phospholipid bilayered micelles for structural studies of polypeptides. Does the ideal bicelle exist? J Magn Reson B 1996;113:267−71.

[93] Sanders C, Prosser R. Bicelles: a model membrane system for all seasons? Structure 1998;6:1227−34.

[94] Warschawski DE, Arnold AA, Beaugrand M, Gravel A, Chartrand É, Marcotte I. Choosing membrane mimetics for NMR structural studies of transmembrane proteins. Biochim Biophys Acta 1808;2011:1957−74.

[95] Prosser RS, Evanics F, Kitevski JL, Al-Abdul-Wahid MS. Current applications of bicelles in NMR studies of membrane-associated amphiphiles and proteins. Biochemistry 2006;45:8453−65.

[96] Popot J-L. Amphipols, nanodiscs, and fluorinated surfactants: three nonconventional approaches to studying membrane proteins in aqueous solutions. Annu Rev Biochem 2010;79:737−75.

[97] Bayburt TH, Sligar SG. Membrane protein assembly into nanodiscs. FEBS Lett 2010;584:1721−7.

[98] Park SH, Berkamp S, Cook GA, Chan MK, Viadiu H, Opella SJ. Nanodiscs versus macrodiscs for NMR of membrane proteins. Biochemistry 2011;50: 8983−5.

[99] Sefcik MD, Schaefer J, Stejskal EO, McKay RA, Ellena JF, Dodd SW, et al. Lipid bilayer dynamics and rhodopsin−lipid interactions: new approach using high-resolution solid-state 13C NMR. Biochem Biophys Res Commun 1983;114: 1048−55.

[100] Marassi FM, Crowell KJ. Hydration-optimized oriented phospholipid bilayer samples for solid-state NMR structural studies of membrane proteins. J Magn Reson 2003;161:64−9.

[101] Fiaux J, Bertelsen EB, Horwich AL, Wüthrich K. NMR analysis of a 900 K GroEL GroES complex. Nature 2002;418:207−11.

[102] Ando T, Kodera N, Takai E, Maruyama D, Saito K, Toda A. A high-speed atomic force microscope for studying biological macromolecules. Proc Natl Acad Sci USA 2001;98:12468−72.

[103] Mari SA, Pessoa J, Altieri S, Hensen U, Thomas L, Morais-Cabral JH, et al. Gating of the MlotiK1 potassium channel involves large rearrangements of the cyclic nucleotide-binding domains. Proc Natl Acad Sci USA 2011;108: 20802−7.

[104] Cavalieri SJ, Bohach GA, Snyder IS. Escherichia coli alpha-hemolysin: characteristics and probable role in pathogenicity. Microbiol Rev 1984;48:326−43.

[105] Orwin PM, Schlievert PM, Dinges MM. Exotoxins of Staphylococcus aureus. Clin Microbiol Rev 2000;13:16.

[106] Bantel H. Alpha-toxin is a mediator of Staphylococcus aureus-induced cell death and activates caspases via the intrinsic death pathway independently of death receptor signaling. J Cell Biol 2001;155:637−48.

[107] Jung M, Vogel N, Köper I. Nanoscale patterning of solid-supported membranes by integrated diffusion barriers. Langmuir 2011;27:7008−15.

[108] Misra N, Martinez J, Huang S, Wang Y, Stroeve P, Grigoropoulos C, et al. Bioelectronic silicon nanowire devices using functional membrane proteins. Proc Natl Acad Sci USA 2009.

[109] Huang S-CJ, Artyukhin AB, Misra N, Martinez JA, Stroeve PA, Grigoropoulos CP, et al. Carbon nanotube transistor controlled by a biological ion pump gate. Nano Lett 2010;10:1812−6.

[110] Menestrina G, Bashford CL, Pasternak CA. Pore-forming toxins—experiments with *S. aureus* alpha-toxin, *C. perfringens* theta-toxin and *E. coli* hemolysin in lipid bilayers, liposomes and intact cells. Toxicon 1990;28:477−91.

[111] Menestrina G. Ionic channels formed by *Staphylococcus aureus* alpha-toxin: voltage-dependent inhibition by divalent and trivalent cations. J Membr Biol 1986;90:177−90.

[112] Menestrina G, Mackman N, Holland IB, Bhakdi S. *Escherichia coli* hemolysin forms voltage-dependent ion channels in lipid membranes. Biochim Biophys Acta 1987;905:109−17.

[113] Prevost G, Mourey L, Colin D, Menestrina G. Staphylococcal pore-forming toxins. Curr Top Microbiol 2001;257:53−83.

[114] Tabard-Cossa V, Trivedi D, Wiggin M, Jetha NN, Marziali A. Noise analysis and reduction in solid-state nanopores. Nanotechnology 2007;18:305505.

[115] McGillivray DJ, Valincius G, Heinrich F, Robertson JWF, Vanderah DJ, Febo-Ayala W, et al. Structure of functional *Staphylococcus aureus* alpha-hemolysin channels in tethered bilayer lipid membranes. Biophys J 2009;96:1547−53.

[116] Kawate T, Gouaux E. Arresting and releasing staphylococcal alpha-hemolysin at intermediate stages of pore formation by engineered disulfide bonds. Protein Sci 2003;12:997−1006.

[117] Krasilnikov OV, Rodrigues CG, Bezrukov SM. Single polymer molecules in a protein nanopore in the limit of a strong polymer−pore attraction. Phys Rev Lett 2006;97:018301.

[118] Bacri L, Oukhaled A, Hémon E, Bassafoula FB, Auvray L, Daniel R. Discrimination of neutral oligosaccharides through a nanopore. Biochem Biophys Res Commun 2011;412:561−4.

[119] Kasianowicz JJ, Brandin E, Branton D, Deamer DW. Characterization of individual polynucleotide molecules using a membrane channel. Proc Natl Acad Sci USA 1996;93:13770−3.

[120] Murphy RJ, Muthukumar M. Threading synthetic polyelectrolytes through protein pores. J Chem Phys 2007;126:051101.

[121] Oukhaled G, Bacri L, Mathe J, Pelta J, Auvray L. Effect of screening on the transport of polyelectrolytes through nanopores. Europhys Lett 2008;82:48003.

[122] Gibrat G, Pastoriza-Gallego M, Thiebot B, Breton M-F, Auvray L, Pelta J. Polyelectrolyte entry and transport through an asymmetric alpha-hemolysin channel. J Phys Chem B 2008;112:14687−91.

[123] Oukhaled A, Biance A-L, Pelta J, Auvray L, Bacri L. Transport of long neutral polymers in the semidilute regime through a protein nanopore. Phys Rev Lett 2012;108:88104.

[124] Oukhaled G, Mathe J, Biance A-L, Bacri L, Betton J-M, Lairez D, et al. Unfolding of proteins and long transient conformations detected by single nanopore recording. Phys Rev Lett 2007;98:158101.

[125] Lathrop DK, Ervin EN, Barrall GA, Keehan MG, Kawano R, Krupka MA, et al. Monitoring the escape of DNA from a nanopore using an alternating current signal. J Am Chem Soc 2010;132:1878−85.

[126] Kang X-F, Gu L-Q, Cheley S, Bayley H. Single protein pores containing molecular adapters at high temperatures. Angew Chem Int Ed 2005;44:1495−9.

[127] Kasianowicz JJ, Burden D, Han L, Cheley S, Bayley H. Genetically engineered metal ion binding sites on the outside of a channel's transmembrane beta-barrel. Biophys J 1999;76:837−45.

[128] Kasianowicz JJ, Burden DL, Han LC, Cheley S, Bayley H. Genetically engineered metal ion binding sites on the outside of a channel's transmembrane beta-barrel. Biophys J 1999;76:837−45.

[129] Gu L-Q, Braha O, Conlan S, Cheley S, Bayley H. Stochastic sensing of organic analytes by a pore-forming protein containing a molecular adapter. Nature 1999;398:686−90.

[130] Rincon-Restrepo M, Mikhailova E, Bayley H, Maglia G. Controlled translocation of individual DNA molecules through protein nanopores with engineered molecular brakes. Nano Lett 2011;11:746−50.

[131] Gu L-Q, Cheley S, Bayley H. Capture of a single molecule in a nanocavity. Science 2001;291:636−40.

[132] Gu L-Q, Bayley H. Interaction of the noncovalent molecular adapter, beta-cyclodextrin, with the staphylococcal alpha-hemolysin pore. Biophys J 2000;79: 1967−75.

[133] Kang X-F, Cheley S, Guan X, Bayley H. Stochastic detection of enantiomers. J Am Chem Soc 2006;128:10684−5.

[134] Banerjee A, Mikhailova E, Cheley S, Gu L-Q, Montoya M, Nagaoka Y, et al. Molecular bases of cyclodextrin adapter interactions with engineered protein nanopores. Proc Natl Acad Sci USA 2010;107:8165−70.

[135] Nguyen TL. Three-dimensional model of the pore form of anthrax protective antigen. Structure and biological implications. J Biomol Struct Dyn 2004;22: 253−65.

[136] Kintzer AF, Thoren KL, Sterling HJ, Dong KC, Feld GK, Tang II, et al. The protective antigen component of anthrax toxin forms functional octameric complexes. J Mol Biol 2009;392:614−29.

[137] Abrami L, Lindsay M, Parton RG, Leppla SH, van der Goot FG. Membrane insertion of anthrax protective antigen and cytoplasmic delivery of lethal factor occur at different stages of the endocytic pathway. J Cell Biol 2004;166:645−51.

[138] Klimpel KR, Molloy SS, Thomas G, Leppla SH. Anthrax toxin protective antigen is activated by a cell surface protease with the sequence specificity and catalytic properties of furin. Proc Natl Acad Sci USA 1992;89:10277−81.

[139] Panchal RG, Halverson KM, Ribot W, Lane D, Kenny T, Abshire TG, et al. Purified *Bacillus anthracis* lethal toxin complex formed *in vitro* and during infection exhibits functional and biological activity. J Biol Chem 2005;280:10834−9.

[140] Karginov VA, Nestorovich EM, Schmidtmann F, Robinson TM, Yohannes A, Fahmi NE, et al. Inhibition of *S. aureus* alpha-hemolysin and *B. anthracis* lethal toxin by beta-cyclodextrin derivatives. Bioorg Med Chem 2007;15:5424−31.

[141] Nestorovich EM, Karginov VA, Berezhkovskii AM, Bezrukov SM. Blockage of anthrax PA63 pore by a multicharged high-affinity toxin inhibitor. Biophys J 2010;99:134−43.

[142] Yan R-X, Chen Z, Zhang Z. Outer membrane proteins can be simply identified using secondary structure element alignment. BMC Bioinf 2011;12:76.

[143] Berven FS, Flikka K, Jensen HB, Eidhammer I. BOMP: a program to predict integral β-barrel outer membrane proteins encoded within genomes of Gram-negative bacteria. Nucleic Acids Res 2004;32:W394−9.

[144] Philippsen A, Im W, Engel A, Schirmer T, Roux B, Muller DJ. Imaging the electrostatic potential of transmembrane channels: atomic probe microscopy of OmpF porin. Biophys J 2002;82:1667−76.

[145] Nestorovich EM, Danelon C, Winterhalter M, Bezrukov S. Designed to penetrate: time-resolved interaction of single antibiotic molecules with bacterial pores. Proc Natl Acad Sci USA 2002;99:9789−94.

[146] Dargent B, Rosenbusch J, Pattus F. Selectivity for maltose and maltodextrins of maltoporin, a pore-forming protein of *E. coli* outer membrane. Biochem Biophys Res Commun 1987;220:136−42.

[147] Kullman L, Winterhalter M, Bezrukov SM. Transport of maltodextrins through maltoporin: a single-channel study. Biophys J 2002;82:803−12.

[148] Schirmer T, Keller T, Wang Y, Rosenbusch J. Structural basis for sugar translocation through maltoporin channels at 3.1 Å resolution. Science 1995;267: 512−4.

[149] Gurnev PA, Oppenheim A, Winterhalter M, Bezrukov SM. Docking of a single phage lambda to its membrane receptor maltoporin as a time-resolved event. J Mol Biol 2006;359:1447−55.

[150] Niederweis M. Mycobacterial porins—new channel proteins in unique outer membranes. Mol Microbiol 2003;49:1167−77.

[151] Mahfoud M, Sukumaran S, Hülsmann P, Grieger K, Niederweis M. Topology of the porin MspA in the outer membrane of *Mycobacterium smegmatis*. J Biol Chem 2006;281:5908−15.

[152] Manrao EA, Derrington IM, Laszlo AH, Langford KW, Hopper MK, Gillgren N, et al. Reading DNA at single-nucleotide resolution with a mutant MspA nanopore and phi29 DNA polymerase. Nat Biotechnol 2012;30:349−53.

[153] Butler TZ, Pavlenok M, Derrington IM, Niederweis M, Gundlach JH. Single-molecule DNA detection with an engineered MspA protein nanopore. Proc Natl Acad Sci USA 2008;105:20647−52.

[154] Wang G, Li X, Wang Z. APD2: the updated antimicrobial peptide database and its application in peptide design. Nucleic Acids Res 2009;37:D933−7.

[155] Cafiso DS. Alamethicin: a peptide model for voltage gating and protein−membrane interactions. Annu Rev Biophys Biomol Struct 1994;23:141−65.

[156] Fox RO, Richards FM. A voltage-gated ion channel model inferred from the crystal-structure of alamethicin at 1.5-Å resolution. Nature 1982;300:325−30.

[157] Bezrukov SM, Vodyanoy I. Probing alamethicin channels with water-soluble polymers effect on conductance of channel states. Biophys J 1993;64:16−25.

[158] Jeon T-J, Malmstadt N, Poulos JL, Schmidt JJ. Black lipid membranes stabilized through substrate conjugation to a hydrogel. Biointerphases 2008;3:FA96−A100.

[159] Boheim G, Hanke W, Eibl H. Lipid phase transition in planar bilayer membrane and its effect on carrier- and pore-mediated ion transport. Proc Natl Acad Sci USA 1980;77:3403−7.

[160] Ketchem R, Hu W, Cross T. High-resolution conformation of gramicidin a in a lipid bilayer by solid-state NMR. Science 1993;261:1457−60.

[161] Cornell B, Braach-Maksvytis V, King L, Osman P, Raguse B, Wieczorek L, et al. A biosensor that uses ion-channel switches. Nature 1997;387:580−3.

[162] Macrae MX, Blake S, Jiang X, Capone R, Estes DJ, Mayer M, et al. Channel platform for detection of phosphatase and protease activity. ACS Nano 2009;3: 3567−80.

Solid-State Nanopore Fabrication

5

Thomas Gibb[1] and Mariam Ayub[2]

[1]*Department of Chemistry, Imperial College London, London;* [2]*Department of Chemistry, University of Oxford, United Kingdom*

CHAPTER OUTLINE

5.1 Introduction

This chapter presents the techniques required to produce a basic nanopore device. Firstly, the materials commonly seen in nanopore devices are presented before being followed with a step-by-step discussion of the fabrication process. Discussion of the fabrication process is centered on a pore constructed within a silicon nitride membrane as, currently, this material is the most commonly seen in the fabrication of inorganic nanopores. Also included are protocols for the key fabrication steps, providing a clear pathway for the reader to produce their own basic nanopore device. While the protocols included in this chapter are not the only means of producing a nanopore device, they allow the production of simple, reproducible devices that can be developed by the reader to suit their own applications.

5.2 Overview of materials

Solid-state nanopores are typically made in robust inorganic materials or organic polymers [1]. Each nanopore "chip" consists of a thin, freestanding membrane fabricated on a support structure most commonly made of silicon. The choice of membrane material is crucial to the nanopore device. The mechanical properties of the freestanding membrane directly affect the ease of the fabrication process and the quality of the final device. The differing electrical properties of the membrane materials also influence the performance of the nanopore while studying single biomolecule translocations. Solid-state materials offer superior chemical and mechanical robustness and can be processed easily using current microfabrication techniques. Most commonly used materials include low stress silicon nitride (SiN_x), SiO_2 [2] SiC [3], Al_2O_3 [4] and graphene [5], which allow the user to precisely control the size of the pore depending on the specific physical and chemical requirements in the measurement.

Silicon-based materials are a popular choice for nanopore fabrication as conventional semiconductor processing techniques allow for the rapid parallel production of multiple chips. Of the silicon-based materials, low-stress SiN_x deposited using low pressure chemical vapour deposition (LPCVD) is most commonly chosen as the membrane material as it is an excellent insulator with high resistivity ($10^{16}\,\Omega$ cm) and dielectric strength (10 MV/cm) [6]. During measurements, the LPCVD SiN_x provides gigaohm-range seal levels, which allows the detection of low ionic currents (tens of picoamperes) through the nanopore. In addition, silicon nitride is chemically stable over a range of temperatures, pH, and electrolyte concentrations. This stability results in SiN_x-based nanopores being extremely versatile with a wide variety of nanopore experiments possible. A further advantage is the similarity of the coefficients of thermal expansion for SiN_x and Si, allowing the deposition of nitride layers on a silicon substrate at high temperatures without large residual stresses at room temperature that may result in cracking of the unsupported nanopore membrane.

In the case of organic polymers, any material with sufficient resistivity and dielectric strength may be used to fabricate the nanopore membrane, as long as it is possible to form sufficiently thin films and etching processes are available. Commonly used materials are polyethylene terephthalate (PET), polycarbonate, and polyimide. Outside of these two families of materials, there are still options for the production of nanopores; mica devices [7], glass nanopipettes [39] and PDMS [40] for example [7].

5.3 **Fabrication methods**

The fabrication of a nanopore is a complex, multistage process. It is necessary to undertake fabrication in a clean room environment to prevent the contamination of devices with external pollutants that reduce the viability of devices to an unworkable level. What follows is the description of a standard procedure for creating a free-standing SiN_x membrane on a silicon wafer followed by a discussion of potential methods of nanopore milling for various systems. The overall fabrication process is highlighted in Figure 5.1.

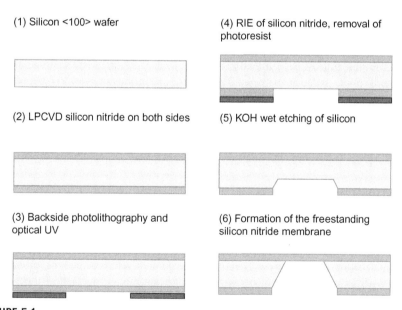

(1) Silicon <100> wafer

(2) LPCVD silicon nitride on both sides

(3) Backside photolithography and optical UV

(4) RIE of silicon nitride, removal of photoresist

(5) KOH wet etching of silicon

(6) Formation of the freestanding silicon nitride membrane

FIGURE 5.1

Main steps of metallic nanopore fabrication process. Steps (1−6) are carried out on whole wafer substrates, the chips are then cleaved, and nanopore milling is done on individual samples. (1) Starting with a double polished Si wafer. (2) SiN_x is deposited on both sides of the Si wafer using LPCVD. (3) Backside lithography is then carried out by exposure to UV for defining areas for SiN_x etch. (4) Subsequently, RIE of SiN_x and removal of photoresist is performed. (5−6) Following this, KOH etching of Si is carried out for the release of the freestanding membrane.

5.3.1 Deposition of silicon nitride films

A prerequisite to making nanopores is the fabrication of solid-state freestanding membranes. SiN_x films are deposited on both sides of a standard 4 in. double-side polished Si $\langle 100 \rangle$ wafers using low-pressure chemical vapor deposition (LPCVD) at a temperature of 825°C using ammonia (NH_3) and dichlorosilane ($SiCl_2H_2$) gases.

$$3SiCl_2H_2 + 4NH_3 \rightarrow Si_3N_4 + 6\,H_2 + 6HCl \qquad (5.1)$$

The deposited film is a product of a chemical reaction between the source gases supplied to the reactor Eq. (5.1). Ensuring only a small amount of residual stress is present in the nitride layer after deposition is essential to allow the formation of freestanding membranes and reduce the possibility of film cracking and spallation. The tensile stress can be reduced by increasing the Si content during LPCVD and depositing nonstoichiometric silicon nitride (SiXNY) films. This is usually done by reducing the $NH_3 : SiH_2Cl_2$ flow rate ratio during the LPCVD process. Furthermore, the thickness of the deposited nitride film can be controlled with subnanometer resolution in the LPCVD process. With thinner membranes exhibiting an increase in the signal to noise ratio. Thinning of the nanopore membrane increases the fragility of the device and thus a balance between sensitivity and usability must be struck when deciding the membrane thickness. Typically, devices with a 20 nm membrane thickness provide a good balance between sensitivity and robustness with membranes below this thickness usually proving too fragile for simple experimentation. Subsequently, multiple membrane templates (generally 50 μm × 50 μm) are then fabricated on the silicon/silicon nitride substrate wafer using standard photolithography techniques.

5.3.2 Photolithography

The deposition of SiN_x onto a Si substrate provides wafers with the correct strata of Si and SiN_x to produce nanopore devices. However, to create the nanopore devices themselves, these layers must be selectively etched away to form the individual nanopore structures. The first stage in this selective etching process is photolithography of a uniform coating of photoresist on the surface of the wafer. For good quality photolithography, the photoresist layer must be highly uniform, which is achieved via a spin coating and baking process. A photoresist is essential in the fabrication process as exposure to a stimulus, in most cases ultraviolet (UV) light, changes its solubility in chemical etchant developers allowing patterns of photoresist to be created on the surface of the wafer. These patterns of photoresist are subsequently used as templates allowing the selective processing of the SiN_x layer.

To form the photoresist patterns shadow masks must first be produced that expose the photoresist to incident light in the correct manner. Shadow masks can selectively protect specific areas of the wafer from any incident light while allowing the exposure of other sections. There are two types of photoresist — positive and negative. The positive allows the user to make a copy of the pattern on the shadow mask whereas the negative will result in an inverse pattern.

After exposure to UV, the photoresist comprises parts which have been chemically altered and parts which remain unchanged. Using developer it is possible to remove either the parts which have or have not been exposed (depending on choice of positive/negative resist).

PROTOCOL: SPIN COATING AND PHOTOLITHOGRAPHY

1. Take a previously processed 4 in. wafer and plasma clean for 10 min.
2. Heat the wafer to 97°C for 5 min.
3. Place the wafer in a spin coater and use a pipette to cover the surface in a layer of AZ 1512 positive photoresist.
4. Spin the wafer at 4000 rpm for 1 min to achieve uniform coverage at a thickness of 1.20 μm.
5. Post spin bake at 115°C for 1 min.
6. Secure the photolithographic mask on top of the photoresist and expose to 420 nm UV radiation at an intensity of 20 MJ/cm^2 for 4 s.
7. Develop in a solution of one part AZ 400 K developer to four parts deionized water for 30 s.
8. Immediately wash with distilled water followed by isopropanol.

A pattern used for preparing freestanding silicon nitride membranes with a positive photoresist is shown in Figure 5.2. Each chip contains photolithographic features suitable for producing a single nitride membrane of 50×50 μm^2. Four of the chips have their dimensions exemplified in Figure 5.2 with a separation of 360 μm between each chip, enabling their separation after processing by breaking the nitride membrane.

PROTOCOL: KOH ETCHING

1. Prepare a solution of 30% wt KOH and heat to 93°C. This solution provides an etch rate of approximately 80 μm/h in the ⟨100⟩ direction.
2. Place the wafer to be etched into the solution with the side to be etched completely exposed to the KOH solution.
3. Ensure there is good flow of the etching solution over the surface of the wafer, for example, with a magnetic stirrer.
4. Observe the etching of the wafer by monitoring the size of the membrane window that has developed.
5. The size of the membrane window should be monitored by removing the wafer from the KOH etching solution, rinsing it with water, and then placing it under an optical microscope.
6. Observation should take place at half hourly intervals until the desired size of membrane window is observed, for the mask as shown in Figure 5.2 50×50 μm^2.

The design of the shadow mask controls the size of the nitride membrane in which the nanopore will be milled. The size of the final nitride membrane is dictated by the size of the photolithographic features of the mask. Due to the anisotropic etching of the silicon, discussed in Section 5.3.3.2, the walls at the edge of the etched pit will be at the angle between ⟨100⟩ and ⟨111⟩ crystallographic directions. Thus, the size of the membrane can be calculated using trigonometry and the thickness of the silicon layer.

$$\text{Membrane dimension} = \text{Photolithographic feature size} - \sqrt{2(\text{depth of silicon})}$$

5.3.3 Etching

After photolithography, etching of the wafer is required to transfer the detailing of the mask onto the wafer itself. Formation of the nanopore structures requires two separate etching steps. The photoresist mask is used to define the areas of

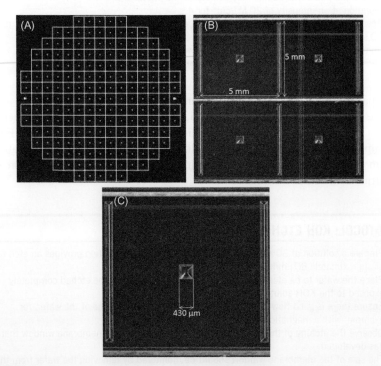

FIGURE 5.2

(A) 4 inch wafer pattern incorporating 175 functional chips. (B) Example of 4 chips (enlarged). Dimensions of each chip shown as 5 mm by 5 mm. (C) Example of a single chip with a central feature size of 430 μm by 430 μm.

silicon nitride for removal via reactive ion etching (RIE) before the nitride layer itself is used as a mask in the subsequent wet chemical etching.

5.3.3.1 Reactive ion etching

SiN_x can be selectively removed from the patterned wafers using a technique known as RIE. In RIE, the wafer substrate is placed inside a reactor, the etching gas is introduced, and the plasma created in the gas mixture using an Radio Frequency (RF) power source. A bias voltage is then applied to accelerate the activated, energetic ions of the plasma toward the surface of the wafer where they react with the nitride layer, etching the material away. Due to the large bias voltage applied between the wafer platter and the upper electrode, the ions of the plasma impinge on the wafer surface with a vector normal to it. This almost universal vertical delivery of ions results in a highly anisotropic etch, essential for creating the well-defined features required for the nitride layer to be used as a mask in subsequent steps (Figure 5.3).

PROTOCOL: TYPICAL CONDITIONS FOR RIE

1. 10 standard cubic centimetres per minute (SCCM) of CF_4.
2. 50 SCCM of argon.
3. 25 W RF power applied.
4. 100 mT pressure inside the reaction chamber.
5. Bias of 40 V applied to the plasma.
6. These conditions produce an etch rate of around 8 nm/min.
7. The etch rate can be increased by increasing the flow rate of CF_4, the power of the RF, the bias voltage, or the pressure inside the reaction chamber.
8. A decrease in the etch rate can be achieved with a decrease in these parameters or an increase in the SCCM of argon.

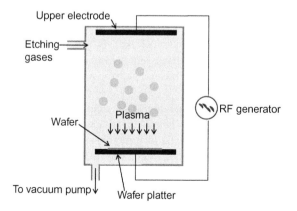

FIGURE 5.3

Diagram of an RIE experimental set up.

The etching gas used in this process can be pure or a mixture of two or more gases. Commonly used etching gases include SF_6 and CF_4 as well as mixtures of these two gases with O_2. All four of these potential gas compositions etch both Si and SiN_x[8], thus the processed wafers' exposure to the plasma must be carefully controlled to prevent excessive removal of material. Exposure times should be calculated by measuring the etch rate of a sample then subsequently using this information to achieve the depth of etch required. Unreactive gases can be added to the etch gas mixture to reduce the reactivity of the plasma and slow the etching rate, enabling greater control of the final etch depth.

5.3.3.2 *Wet etching*

After the process of RIE, the SiN_x layer is used as a mask to enable the selective removal of the silicon wafer to produce the freestanding membranes required for nanopore fabrication. Etching of the exposed Si substrate should be undertaken using an anisotropic KOH etch. KOH is highly selective for Si [8]. The etch selectivity for Si vs. LPCVD SiN_x is over 10^4:1 at typical etching conditions of 7 M KOH at 80°C, allowing the formation of freestanding nitride membranes [6].

Using a KOH etching solution results in a highly anisotropic etch as the removal of Si is 400 times faster in the $\langle 100 \rangle$ and $\langle 110 \rangle$ directions than in the $\langle 111 \rangle$ directions because of the crystal structure of the silicon [6]. Due to this anisotropy, pyramidal pits etch down to the Si wafer in areas that are not protected by SiN_x. The rate of the KOH etch depends on both the concentration and the temperature of the solution, with concentrations below 30% wt KOH resulting in rough etches [9]. The growth of these pits should be monitored over several hours via a light microscope until the correct membrane size as determined by the photolithographic feature size is observed. At this point, the etch should be terminated by washing the wafer with deionized water followed by isopropanol. The process is detailed in Figure 5.4.

5.3.4 Pore milling

The final step in the production of the nanopore membrane is the milling of the nanopore itself. This process can be undertaken in many different ways with the most commonly seen approaches discussed below: nonfocused and focused ion beams (FIBs), high-power electron beams, and for polymer membranes, ion track etching.

The first pores successfully used for DNA analysis, developed by Li et al. [11], were produced using an ion beam technique. In their approach, a bowl-shaped cavity was created on one side of a membrane using RIE before an energetic beam of Ar^+ ions was used to sputter away the membrane from the reverse side. The sputtering process involves a cascade of elastic collisions as the incident ion impinges on the surface of the SiN_x. Atoms are ejected or sputtered from the substrate surface if the incident ion has transferred a sufficient component of its kinetic energy to the surface atom to allow it to overcome the surface binding energy.

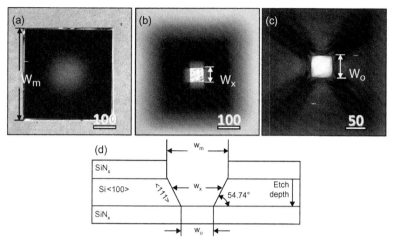

FIGURE 5.4

Anisotropic etching of $\langle 100 \rangle$ Si. (A) Measurement of RIE opened window, W_m. (B) Snapshots of the intermediate etched window are taken during the etch and used to calculate the etch rate, given by $\tau_{etch} = (W_m - W_x)/\sqrt{2}t$, where t is time. (C) Fully etched chip with a membrane dimension of W_o. (D) Overall schematic.

Source: Adapted from Ref. [10].

As sputtering removes material from the nitride surface, the planar surface will eventually intercept the bowl-shaped cavity previously formed with RIE. When this intersection occurs, a nanopore is formed, detected by the leakage of the incident Ar^+ ions through the membrane. Creating a molecular-sized aperture requires careful control of the sputtering process, usually involving a feedback system that detects the ions transmitted through the opening pore, stopping the erosion of the membrane when the number of detected ions corresponds to the required pore diameter. This method has been used for the successful fabrication of pores down to 1.5 nm in diameter.

The ion beam approach used by Li et al. [11] to form nanopores has been developed to include the use of an FIB. FIBs are commonly used in imaging and dissecting semiconductor devices and can be used to drill nanopores in freestanding membranes, providing flexibility with regard to pore size and the membrane material used. In principle, any ion with the energy to sputter material away from the membrane surface can be used for nanopore fabrication with the most commonly seen ions being Ga^+, He^+, Ne^+, Ar^+, Kr^+, and Xe^+[12]. Ga^+ ions are the most frequently used for nanopore milling, despite the potential for ion implantation into the membrane and the associated modification of the nanopore's surface charge.

The milling of pores in the sub 5 nm range with an FIB was pioneered by Gierak et al. [3]. Here, a 35 keV Ga^+ beam was focused with an electric field to form a 5 nm full width at half maximum probe and used to drill pores through a

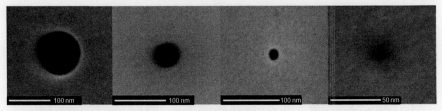

FIGURE 5.5

Solid-state nanopores prepared via a FIB in SiN$_x$. Four different nanopores were fabricated using an FIB. Each hole was first etched through a SiN membrane by the direct application of a focused beam before the diameter was reduced by scanning the ion beam across the pore and its surrounding area [13].

20 nm SiC membrane. In this process, the Ga$^+$ ions of the focused beam are heavy and energetic enough to result in highly localized sputtering, with no requirement for a previously etched cavity on the reverse of the membrane to create a nanometer scale pore. Subsequent scanning of the area around the pore leads to controlled shrinkage of the pore and its aperture becoming more circular [13,14] (Figure 5.5).

PROTOCOL: FIB MILLING

1. Dip the device into isopropanol then to hot deionized water repeatedly to remove organic contaminants.
2. Plasma clean the membrane for 5–10 min to remove the final traces of organic impurities.
3. Mount the membrane inside the FIB with carbon tape or silver paint to ground the device. Grounding of the device prevents the SiN$_x$ from becoming positively charged during milling. A charged SiN$_x$ surface is difficult to image, preventing the ion beam from being properly focused, resulting in undesirably large nanopores.
4. Set the magnification, beam current, and beam energy to values that give the minimum spot size for the FIB used.
5. For a set beam current and minimum spot size, the ion dose (ions/area) and ion flux (dose/time) can be varied by changing the milling time.
6. A dose test should then be carried out to find the optimal milling time. This is done by milling adjacent pores for different times, then imaging the pores to find the milling time, which results in a pore with the desired parameters (Figure 5.6).
7. In general, to drill through 100 nm SiN$_x$ doses should be around 3×10^{23} cm^{-2} and milling times below 10 s to achieve sub-100 nm pores. These values are only approximate; exact values depend on the quality of beam focus, the angle of the incident ions, and density of the nitride layer among other variables.

The FIB milling approach to nanopore milling has been further developed with the introduction of the helium ion microscope (HIM). The HIM uses an ion source with an atomically sized tip allowing the production of a highly focused ion beam and the fabrication of pores with diameters as low as 4 nm [15].

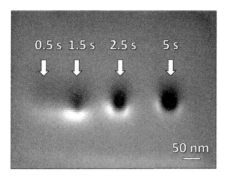

FIGURE 5.6

FIB dose test. A typical FIB dose test is used to optimize nanopore milling.

Source: Adapted from Ref. [10].

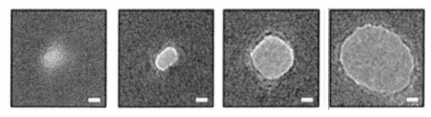

FIGURE 5.7

Solid-state nanopores prepared via a HIM in SiN. Four different nanopores fabricated using a HIM and imaged using TEM. Scale bars are 5 nm.

Source: Adapted from Ref. [15].

In addition, this technique has several advantages compared to traditional gallium ion milling in which it offers a higher resolution, direct thinning of the membrane, and the absence of ion implantation [16] (Figure 5.7).

An alternative to the use of ion beams in nanopore milling is the use of high-energy electron beams. The use of high-energy electron beams allows the production of solid-state nanopores with single nanometer precision using a commercial transmission electron microscope (TEM) [2]. Here, the process begins with the creation of a large, up to 50 nm, pore in an amorphous SiO_2 or SiN_x [17] layer that is systematically shrunk with the application of an electron beam with intensity in the range of $10^5 - 10^7$ A/m^2. Under these conditions, pore shrinkage is seen at nanometer per minute rates, slow enough to allow complete control of nanopore diameter. The final nanopore diameter is only limited by the resolution of the microscope and the surface roughness of the membrane.

Nanopore shrinkage occurs under electron irradiation of these intensities as viscous flow is induced in the amorphous silicon oxide [18] or nitride layer. The shrinkage of the nanopore is then driven by a reduction in surface energy, F.

The change in free energy due to a pore compared to an intact sheet of the membrane material is:

$$\Delta F = 2\pi\gamma(rh - r^2) \tag{5.2}$$

where γ is the surface tension, r the pore radius, and h the membrane thickness. It can be seen that pores with a radius $r < h/2$ can lower their surface energy by reducing their size, applicable at any length scale [2] (Figure 5.8).

5.3.5 Ion track etching

Where the membrane film is a polymer rather than an oxide or nitride layer, the track-etching technique is an alternative to the methods already discussed. Track etching is mainly employed in PET [19], polycarbonate, and polyimide films, as well as in mica devices [20]. In this process, a commercially available dielectric film with a thickness in the order of $10\,\mu m$ is first irradiated with energetic heavy ions. These ions travel through the membrane creating zones

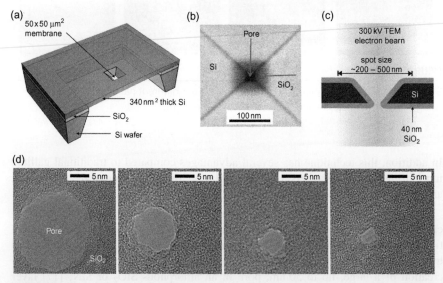

FIGURE 5.8

(A) Cross-section view of a typical device. It consists of a 340-nm-thick free-standing single-crystalline silicon membrane, supported by a KOH-etched Si wafer. (B) Top-view SEM image of a 20 nm nanopore after thermal oxidation, surrounded by a SiO_2 layer of ~ 40 nm thickness. (C) Cross-sectional view of the pore inside the electron microscope. (D) Sequence of images obtained during imaging of a silicon oxide pore in a TEM microscope. The electron irradiation causes the pore to shrink gradually to a size of ~ 3 nm [2].

of local damage. These pathways of damage through the film are commonly known as latent tracks and are subsequently etched to obtain pores with the required dimensions.

Typically, tracks are generated with ions of Xe, Pb, Au, or U using energies in the GeV range. Energies of this magnitude allow the ions to penetrate a $\sim 100 \, \mu m$ polymer material, allowing the creation of nanopore by wet chemical etching. Uniquely, the track-etching technique results in a direct correlation between the number of heavy ions passing through the membrane and the number of pores created during the etching process. Simple fabrication of films with single nanopores was enabled by technical advances in the irradiation step that allowed the exposure of each membrane to only one heavy ion [21] (Figure 5.9).

5.4 **Control of pore size**

One of the most crucial aspects of nanopore production is obtaining a high degree of control over the size of the pore. Controlling the pore size of these devices is important to ensure adequate sensitivity and reproducibility of devices and thus the results derived from them. Each individual method of producing nanopores has separate variables that must be controlled when tuning the pore size, as discussed in turn below.

5.4.1 **FIB milling**

When tuning the size of nanopore produced in the FIB milling process, the quality of the ion beam itself is the first and most obvious variable to consider. To drill small pores capable of single molecule detection, the ion beam must have the smallest diameter possible at the point where it interacts with the membrane. This should be achieved via optimization of variables such as astigmatism and the use of large magnifications to focus the ion beam on the smallest area possible.

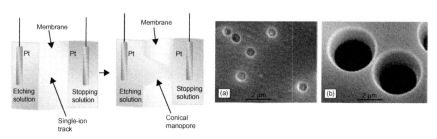

FIGURE 5.9

Synthetic nanopores with a high aspect ratio. Conical nanopores prepared in 12 mm thick foils of PET (left) and polyimide using the track-etching technique (right) [19].

During irradiation by Ga^+ ions, the SiN_x may become charged due to its insulating nature resulting in an accurate focus on the substrate becoming hard to achieve and a loss in nanopore milling precision. To prevent focusing problems, each device should be grounded with the use of carbon tape or conductive silver paint to hold the sample in place.

The beam quality and focus are not the only things that have to be considered, when milling nanopores with an FIB setup. The ability of an FIB to mill quickly through an arbitrary material is derived from the sputtering mechanism by which atoms are ejected from the surface and the near surface of the target [22]. The sputter yield, defined as the number of atoms ejected per incident ion, is high (typically >2) for a commonly used 30 keV Ga^+ ion beam impinging on most materials. A high yield is seen as the incident ion that can transfer much of its kinetic energy to a target atom, overcoming the surface binding energy, E_s, or the bulk displacement energy, E_d. Modeling the interaction between an incident ion and a target atom as an elastic two-body collision, the maximum transferable energy, E_{max}, is found to be [23]:

$$E_{max} = \frac{4M_1M_2}{(M_1M_2)^2}E_0 \tag{5.3}$$

where E_0 is the kinetic energy of the incident particle, and M_1 and M_2 are the masses of the incident ion and the target atom, respectively. Equation (5.3) indicates that the incident kinetic energy can be most efficiently transferred to the target when the masses of the incident and target atoms are closely matched. Nevertheless, hundreds to thousands of electronvolts are transferred in typical collisions with 30 keV Ga^+ ions, even for targets with dissimilar masses.

Although the sputtering process makes FIB milling effective, it also limits the spatial resolution that can be achieved. The 30 keV energy of incident Ga^+ ions from an FIB greatly exceeded E_s and E_d which typically consists of several electron volts. A single incident ion therefore generates a cascade of collisions within the target in which many atoms become displaced or sputtered. The lateral extent of the collision cascade is typically about 10 nm; thus milling structures with critical dimensions in the single nanometer range can become extremely difficult.

The effect of the collision cascade on nanopore diameter can be mitigated by optimizing the milling protocol to expose the membrane to the ion beam for the shortest possible duration. The exposure time can be optimized by determining the milling rate of the ion beam through the substrate and this information is used alongside the membrane thickness to determine the minimum time for a pore to be milled. Using such a protocol will ensure that as few atoms as possible are sputtered by the beam, thus achieving the minimum possible nanopore broadening due to unavoidable collision cascades. The effect of the FIB exposure time on the nanopore diameter is illustrated in Figure 5.10.

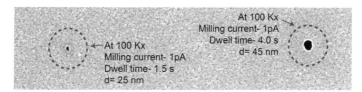

FIGURE 5.10

SEM images of nanopores milled using the FIB technique. The image is of two different pores milled through a Au/SiN$_x$ membrane. The effect of a longer exposure to the FIB is clearly demonstrated. Magnification of the image is set to 30,000 times.

5.4.2 Scanning Transmission Electron Microscopy (STEM) milling

Controlling the size of nanopores made by the STEM technique differs to that of other techniques as here we are controlling the amount of pore shrinkage rather than growth. The controlled shrinking of pores with an electron beam is simply a progression toward a thermodynamic minimum. It is the electron beam that liquefies the amorphous surface layer and thus provides the energy input that allows the kinetic barriers to pore shrinkage to be overcome. By lowering the beam intensity or shutting it off, the shrinking process can be stopped within seconds. This combined with a linear rate of pore shrinkage [2] results in the facile control of pore size on the nanometer scale. If favored, the shrinking process can be accelerated by at least an order of magnitude by increasing the electron intensity before the intensity is lowered to achieve ultimate control. Nanopore size can be monitored during processing using the same electron beam as used for shrinkage providing accessible, *in situ* monitoring.

5.4.3 Ion track etching

When ion track etching is used to produce nanopores, it is the wet etching step, after the bombardment of the membrane by heavy ions, which controls the size of the nanopores. The wet etching process is highly controllable with key variables in the control of pore diameter being the etchant concentration, the etching temperature, and the time duration of the etching process. These three variables must be tuned in unison to achieve the desired pore diameter, altering the track to bulk etch rate ratio and the time over which the pore is allowed to develop. In order to obtain single nanometer-sized pores, the etching step is undertaken in a conductivity cell to simultaneously infer the diameter of the pore from the equation [24]:

$$G = \sigma \left[\frac{4l}{\pi d^2} + \frac{1}{d} \right]^{-1} \tag{5.4}$$

where l is the pore length, d the pore diameter, σ the conductivity of the etching media, and G the conductance of the pore.

5.5 Surface modification

While the process used to fabricate the nanopore can be used to control its dimensions, it is also possible to use a multitude of other techniques to tune this crucial parameter. These techniques are based around modifying the surface of the pore and the surrounding areas with metallic, inorganic, organic, or biological materials, providing control over the pore size as well as the ability to alter the chemical and physical properties of the pore itself.

5.5.1 Inorganic and metallic materials

The ease of deposition and inherent stability of metallic and inorganic materials results in their frequent use to modify the surface of a pore after fabrication. Atomic layer deposition (ALD) is commonly used to deposit inorganic materials onto the surface of a pore with subnanometer precision. Single layers of high dielectric constant materials such as Al_2O_3[4,25] may be deposited onto a prefabricated nanopore membrane, improving the performance of the device by simultaneously reducing the pore diameter, and enhancing the device's mechanical properties and noise performance. The use of ALD to facilitate the modification of the pore surface has been further advanced with the use of multilayered nanolaminates allowing more complex, three terminal graphene nanopore devices [26].

Other, less complex alternatives to the use of ALD for the surface modification of nanopores with metals are sputtering and evaporation. The deposition of a thin titanium adhesion layer followed by gold has been shown to be successful in reducing a nanopore's diameter [27]. The deposited metallic layers conform to the surface of the pore, allowing a controlled reduction in nanopore size. Electrodeposition can be employed to further control the pore dimensions. Here, *in situ* measurement of the nanopore's ionic conductance can be used to provide a real-time measurement of the pore's diameter during deposition [28].

While the techniques listed above successfully modify the surface of the nanopore itself; their effects are not localized to this area alone. Techniques such as these result in the entire surface of the membrane being coated with the material used to modify the pore, in turn ensuring that the properties of the entire membrane are altered. This limitation can, however, be circumvented via local oxide deposition, a process in which the deposition of a precursor is induced locally to the nanopore by exposure to either an ion [29] or an electron beam [30].

5.5.2 Silane and thiol-based modifications

One method of modifying the surface is through employing organosilane chemistry. The silanizaion of the nanopore surface is achieved via either a vapor or solvent based deposition. To covalently bind to the membrane surface, the silane molecules require exposed hydroxyl groups (which can be achieved by treating the nanopore in boiling piranha solution (3:1 mixture of H_2SO_4 and H_2O_2).

In solution, silicon nitride is known to acquire a negative surface charge density of -0.02 C m^{-2}, at neutral pH [41] owing to the deprotonation of native silanol groups. Dressing the pore surface with a variety of organic coatings not only makes it more biologically friendly but further allows control of surface charge, hydrophobicity, and chemical functionality.

To achieve a durable, high density silane film, a post-silanization bake at approximately 100°C for 16 h is often considered optimal. [31]. This step results in the cross-linking of the organosilane molecules at any unreacted alkoxy groups on the surface, although care must be taken to preserve the reactivity of the molecule's tail group. The surface modification of nanopores with organosilanes is popular as they provide a convenient way to affix functional groups to the surface of the membrane via a robust covalent linkage. The wide variety of tail groups available results in the potential for many different surface chemistries at the nanopore.

Not all nanopores can be modified with silanes. Nanopores that have been fabricated or previously modified with gold are often treated with thiol terminated self-assembled monolayers (SAMs) [32]. The self-assembly of these thiol layers results in the relatively easy modification of the nanopore with the properties of the surface modification determined by the SAM tail groups in a similar fashion to the organosilane layer.

Such modifications have previously been used to modulate surface charge, provide steric effects around the nanopore that confer molecular selectivity, or enable a pore to respond to environmental conditions. The nanopore diameter can also be effectively tuned using the molecules by altering the length of the central spacer between the head and the tail groups. The potential for nanopore modification using these methods has been explored by both Wanunu and Meller [33] and Hou et al. [34].

5.5.3 Biological modifications

Often, nanopores are used as biological sensors; thus the ability to modify the pore surfaces with biological molecules is key. These biological modifications can enhance and tailor the properties of solid-state nanopores for sensing. The surfaces can be modified to remove undesired properties. For example, solid-state pores exhibit enhanced low frequency conductance fluctuations, which affect the quality of the recordings. This problem can be mitigated e.g. by chemical modification of the channel walls using chemical vapor deposition. Many biological molecules can be used to modify the nanopore surface including: lipids, proteins, and nucleic acids for various applications.

The surface modification of nanopores with nucleic acids and proteins usually relies on the prior treatment of the surface with a thiol or organosilane layer [35,36]. An additional aim for pore modification is to emulate the ion selectivity of some biological channels by mimicking their natural design principles.

In contrast to proteins and nucleic acids, lipids can often be added directly to the membrane surface to form a supported bilayer [37]. Adjusting the lipid composition of the bilayer allows the nanopore diameter and surface chemistry to be tuned with careful choice of tail length and functional group. Recognition elements can also be introduced to the bilayer, aiding analyte detection by increasing their dwell time. Furthermore, lipid bilayer coatings have proved useful in the analysis of species that have a tendency to clog nanopores such as amyloid-β-aggregates [38]. Importantly, incorporating biological molecules into a solid-state device opens up avenues towards the creation of wafer-scale parallel device arrays that may be useful for genomic and protein sensing.

References

[1] Dekker C. Solid-state nanopores. Nat Nanotechnol 2007;2:209−15.

[2] Storm AJ, Chen JH, Ling XS, Zandbergen HW, Dekker C. Fabrication of solid-state nanopores with single-nanometre precision. Nat Mater 2003;2:537−40.

[3] Gierak J, Madouri A, Biance A, Bourhis E, Patriarche G, Ulysse C, et al. Sub − 5 nm FIB direct patterning of nanodevices. Microelectron Eng 2007;84:779−83.

[4] Venkatesan BM, Shah AB, Zuo JM, Bashir R. DNA sensing using nanocrystalline surface-enhanced Al_2O_3 nanopore sensors. Adv Funct Mater 2010;20: 1266−75.

[5] Fischbein MD, Drndić M. Electron beam nanosculpting of suspended graphene sheets. Appl Phys Lett 2008;93:113107.

[6] Gad-el-Hak M. MEMS: design and fabrication. New York: CRC Press; 2005.

[7] Fleischer RL, Price PB, Walker RM. Nuclear tracks in solids: principles and applications, 605. Berkeley, CA: University of California Press; 1975.

[8] Williams KR, Gupta K, Wasilik M. Etch rates for micromachining processing—part II. J Microelectromech Syst 2003;12:761−78.

[9] Seidel H, Csepregi L, Heuberger A, Baumgärtel H. Anisotropic etching of crystalline silicon in alkaline solutions. J Electrochem Soc 1990;137:3612.

[10] Ivanov AP. Novel nanopore/nanoelectrode architectures for biomolecular analysis; Imperial College London; 2012.

[11] Li J, Stein D, McMullan C, Branton D, Aziz MJ, Golovchenko J, et al. Ion-beam sculpting at nanometre length scales. Nature 2001;412:166−9.

[12] Cai Q, Ledden B, Krueger E, Golovchenko JA, Li J. Nanopore sculpting with noble gas ions. J Appl Phys 2006;100:3264−7.

[13] Lo CJ, Aref T, Bezryadin A. Fabrication of symmetric sub-5 nm nanopores using focused ion and electron beams. Nanotechnology 2006;17:3264−7.

[14] Stein D, Li J, Golovchenko J. Ion-beam sculpting time scales. Phys Rev Lett 2002;89:2−5.

[15] Yang J, Ferranti D, Stern L, Sanford C, Huang J, Ren Z, et al. Rapid and precise scanning helium ion microscope milling of solid-state nanopores for biomolecule detection. Nanotechnology 2011;22:285−310.

[16] Marshall MM, Yang J, Hall AR. Direct and transmission milling of suspended silicon nitride membranes with a focused helium ion beam. Scanning 2012;34:101−6.

[17] Kim MJ, McNally B, Murata K, Meller A. Characteristics of solid-state nanometre pores fabricated using a transmission electron microscope. Nanotechnology 2007;18:205–302.

[18] Ajayan PM, Iijima S. Electron-beam-enhanced flow and instability in amorphous silica fibres and tips. Philos Mag Lett 1992;65:43–8.

[19] Siwy Z, Apel P, Baur D, Dobrev D, Korchev Y, Neumann R, et al. Preparation of synthetic nanopores with transport properties analogous to biological channels. Surf Sci 2003;532–535:1061–6.

[20] Howorka S, Siwy ZS. Nanopores: generation, engineering and single-molecule applications. In: Peter Hinterdorfer, van Oijen AM, editors. Handbook of single-molecule biophysics. New York: Springer; 2009. pp.293–339.

[21] Spohr R. Method for producing nuclear traces or microholes originating from nuclear traces of an individual ion. International Classification: H01J 2300, Patent number: 4369370, 1983.

[22] Orloff JLS, Utlaut MW. High resolution focused ion beams: FIB and applications. New York: Springer; 2005.

[23] Landau LD, Lifshitz EM. Mechanics. Butterworth-Heinemann: Oxford; 1976.

[24] Kowalczyk SW, Grosberg AY, Rabin Y, Dekker C. Modeling the conductance and DNA blockade of solid-state nanopores. Nanotechnology 2011;22:315101.

[25] Chen P, Mitsui T, Farmer D, Golovchenko J, Gordon R, Branton D, et al. Atomic layer deposition to fine-tune the surface properties and diameters of fabricated nanopores. Nano Lett 2004;4:1333–7.

[26] Venkatesan BM, Estrada D, Banerjee S, Jin X, Dorgan V, Bae M-H, et al. Stacked graphene-Al_2O_3 nanopore sensors for sensitive detection of DNA and DNA–protein complexes. ACS Nano 2012;6:441–50.

[27] Wei R, Pedone D, Zürner A, Döblinger M, Rant U. Fabrication of metallized nanopores in silicon nitride membranes for single-molecule sensing. Small 2010;6:1406–14.

[28] Ayub M, Ivanov A, Hong J, Kuhn P, Instuli E, Edel J, et al. Precise electrochemical fabrication of sub − 20 nm solid-state nanopores for single-molecule biosensing. J Phys Condens Matter 2010;22:454128.

[29] Nilsson J, Lee JRI, Ratto TV, Létant SE. Localized functionalization of single nanopores. Adv Mater 2006;18:427–31.

[30] Danelon C, Santschi C, Brugger J, Vogel H. Fabrication and functionalization of nanochannels by electron-beam-induced silicon oxide deposition. Langmuir 2006;22:10711–5.

[31] Halliwell CM, Cass AE. A factorial analysis of silanization conditions for the immobilization of oligonucleotides on glass surfaces. Anal Chem 2001;73:2476–83.

[32] Vericat C, Vela ME, Benitez G, Carro P, Salvarezza RC. Self-assembled monolayers of thiols and dithiols on gold: new challenges for a well-known system. Chem Soc Rev 2010;39:1805–34.

[33] Wanunu M, Meller A. Chemically modified solid-state nanopores. Nano Lett 2007;7:1580–5.

[34] Hou X, Guo W, Jiang L. Biomimetic smart nanopores and nanochannels. Chem Soc Rev 2011;40:2385–401.

[35] Wei R, Gatterdam V, Wieneke R, Tampé R, Rant U. Stochastic sensing of proteins with receptor-modified solid-state nanopores. Nat Nanotechnol 2012;7:257–63.

[36] Mussi V, Fanzio P, Repetto L, Firpo G, Stigliani S, Tonini GP, et al. "DNA-Dressed Nanopore" for complementary sequence detection. Biosens Bioelectron 2011;29:125−31.

[37] Yusko EC, Johnson JM, Majd S, Prangkio P, Rollings RC, Li J, et al. Controlling protein translocation through nanopores with bio-inspired fluid walls. Nat Nanotechnol 2011;6:253−60.

[38] Yusko EC, Prangkio P, Sept D, Rollings R, Li J, Mayer M, et al. Single-particle characterization of Aβ oligomers in solution. ACS Nano 2012;6:5909−19.

[39] Karhanek M, Kemp J, Pourmand N, Davis R, Webb C. Single DNA molecule detection using nanopipettes and nanoparticles. Nano Lett 2005;5:403−7.

[40] Saleh OA, Sohn LL. An artificial nanopore for molecular sensing. Nano Lett 2003;3:37−8.

[41] Sonnefeld J. Determination of surface charge density parameters of silicon nitride. Colloids Surf A: Physicochem Eng Asp 1996;108(1):27−31.

Case Studies Using Solid-State Pores

Gaurav Goyal[1], Kevin J. Freedman[2], Anmiv S. Prabhu[1] and Min Jun Kim[1,3]

[1]School of Biomedical Engineering, Science and Health Systems, Drexel University, Philadelphia, PA; [2]Department of Chemical and Biological Engineering, Drexel University, Philadelphia, PA; [3]Department of Mechanical Engineering and Mechanics, Drexel University, Philadelphia, PA

CHAPTER OUTLINE

6.1 Introduction

Biological nanopores have been widely used for studying DNA and RNA translocations, protein binding, and many other interesting applications as discussed in Chapter 5; however, the pores and the lipid bilayer in which they are suspended suffer from limitations like fixed pore diameter and length, mechanical instability, and can operate over very limited ranges of pH and voltages. These limitations of biological nanopores have been addressed by the use of solid-state nanopores which are artificially drilled holes in silicon nitride (or silicon oxide or graphene) membranes. The solid-state technology makes it possible to fabricate robust nanopores with variable pore dimensions (discussed in this chapter) which can be used over a much wider range of experimental conditions. These solid-state pores have perfectly complemented the biological nanopores for single molecule detection and analysis by expanding the experimental repertoire and by incorporating new electrical and/or optical detection strategies.

This chapter will focus on various applications for which solid-state nanopores have been used. We will start with DNA and RNA translocations and then move to DNA unzipping, optical detection, DNA sequencing, and finally end with protein analysis using the solid-state nanopores.

6.2 DNA and RNA translocations

DNA and RNA translocation studies have been the focus for the majority of nanopore experiments to date. Advances in nanoscale manufacturing in the past decade have enabled fabrication of size-controlled solid-state nanopores, thereby expanding the possibilities for DNA analysis offered by the biological pores. The use of solid-state pores for DNA analysis started with the seminal report from Golovchenko's group at Harvard University in 2001 [1]. They reported that the feedback-controlled ion sculpting technique could be used to drill holes in silicon nitride (SiN) membranes with nanometer scale precision. These size-controlled nanopores were proposed to be used for the development of single molecule sensing devices and authors demonstrated detection of double-stranded DNA (ds-DNA) passing through a 5 nm diameter pore. They also put forward the idea that these nanopore-based devices are like molecular microscopes and can be used to probe individual biomolecules in their native state with unprecedented resolution. And because solid-state technology enables us to produce pores of different dimensions, these pores can be exchanged in the sensing device as one would change lenses on a microscope [2]. The device they used consisted of the electrolyte (potassium chloride, KCl) filled polydimethylsiloxane (PDMS) chambers separated by a SiN membrane having a single nanopore (Figure 6.1). The ds-DNA was introduced in the *cis* chamber and a voltage bias was applied using Ag/AgCl electrodes. When the *cis* chamber was connected to the ground and the *trans* chamber was positively biased, the negatively charged DNA moved across the pore into the *trans* chamber. The movement of DNA molecules across the pore transiently displaced ions flowing through it, resulting in molecule-induced quantized resistive pulses. These spikes could not be observed before adding the DNA or when the polarity was reversed. In Figure 6.1C, I_o represents the open pore current when no DNA is passing through the pore; I_b and t_d are the magnitude and time duration of the current blockades, respectively. In addition to the simple events (like the one shown in Figure 6.1C), authors also reported several complex events corresponding to different folding conformations of DNA. As the DNA molecules folded on themselves and passed through the nanopore, it increased the magnitude of the current blockade for that part of the event.

Similar results demonstrating folded conformations of ds-DNA passing through a 15 nm pore were also reported by Chen et al. [3]. They observed seven

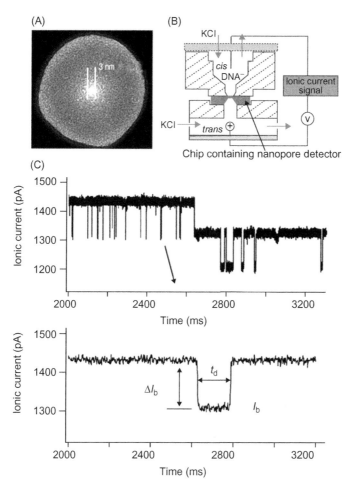

FIGURE 6.1

Experimental setup by Li et al. [2] (A) 3 nm diameter pore drilled by feedback-controlled ion sculpting technique, (B) assembled device with the nanopore chip separating *cis* and *trans* chambers, and (C) current spikes observed upon adding the DNA.

Source: Adapted with permission from Ref. [2]. Copyright (2003) Macmillan Publishers Ltd.

different current signatures indicating different folding conformations of the DNA. Figure 6.2A and B compares the current signature for linear and folded DNA and Figure 6.2C shows seven different current waveforms recorded by the authors. It was suggested that DNA stretches and becomes linear near the nanopore because of inhomogeneous electric field. This stretching was observed to be a function of voltage bias and more linear DNA translocations were registered at higher voltages. The nanopores used in this study were modified by depositing an

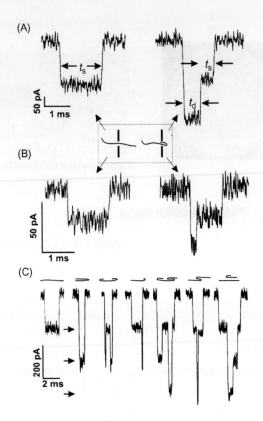

FIGURE 6.2

Translocation events recorded by Chen et al. [3] (A) and (B) linear versus folded DNA passing through the nanopore. Left panel images represent current traces for linear DNA and the right panel for folded DNA, t_s and t_d are event duration for linear DNA and doubly folded DNA respectively, and (C) seven different current traces recorded by the authors.

Source: Adapted with permission from Ref. [3]. Copyright (2004) American Chemical Society.

atomic layer of Al_2O_3, which reduced electrical noise of SiN membranes and enhanced DNA capture rate.

Nanopores fabricated in SiN membranes have also been used for sizing the ds-DNA molecules. Different translocation times were observed for 100 base pair (bp), 600 bp, and 1500 bp length samples when they were driven through a 2.4 nm diameter pore created in 30 nm thick SiN membrane [4]. Correlation of DNA length and translocation time was also studied by Storm et al. [5] using 10 nm diameter pores in 20 nm thick silicon oxide membranes. They analyzed translocation of DNA in size range of 6.5−97 kbp and discovered that event duration increased exponentially with the length of the DNA. In another set of experiments, they also detected DNA passing through the pore in linear and folded states and estimated the folding positions in DNA [6].

These reports on detection and sizing of the ds-DNA and analysis of its different folding conformations presented a new opportunity to study individual DNA molecules. Before the use of solid-state pores, similar analysis for the ds-DNA molecules was not possible because ds-DNA (cross section 2.4 nm) could not translocate through the fixed aperture (1.5 nm) of the biological pores. The works discussed above established the merit and versatility of solid-state nanopores and encouraged many other research groups to use this sensing platform to study individual molecular details of biomolecules and explore the possibility of DNA sequencing using these devices.

After these successful attempts to detect ds-DNA, the focus shifted onto the physics of translocation and DNA−pore interaction in order to exploit the full range of experimental possibilities using the solid-state nanopores. The work by Fologea et al. [7,8] used various experimental conditions and demonstrated the robustness and durability of solid-state nanopores. They explored the effects of electrolyte concentration, pH, viscosity, temperature, and the applied voltage on ds-DNA translocation. They observed a significant decrease in translocation time when different concentrations of glycerol (0−50%) were added to KCl to increase the electrolyte viscosity. Similarly, a twofold increase in translocation time was observed when experiments using 20% glycerol were performed at 4°C instead of at room temperature. Using a combination of these experimental parameters, translocation speed of DNA could be brought down from 30 to 3 bases/μs. However, this strategy to reduce the translocation speed also reduced the ion mobility and event amplitude, resulting in a low signal-to-noise ratio (SNR). Attempts were also made to slow down DNA speed by functionalizing the nanopores with amino-silane to create a positively charged surface which transiently interacted with negatively charged DNA [9]. Recently, it has been achieved by using lithium chloride (LiCl) as the electrolyte instead of KCl [10]. The binding of lithium ions screens the negative charge on the DNA molecules, thereby lowering the translocation speed. The effectiveness of a counter ion (Li^+ or K^+) in lowering the charge on DNA is dependent on the size of the ion. As illustrated in Figure 6.3, the higher ionic strength of LiCl increases both the event amplitude and the translocation time resulting in high SNR. Using 4 M LiCl, Kowalczyk et al. [10] achieved a 10-fold decrease in translocation speed of ds-DNA passing through a 20 nm pore.

These are significant results in the direction of developing nanopore-based rapid DNA sequencers. As shall be discussed in later sections of this chapter, in order to achieve single base resolution for DNA sequencing, it is imperative to slow down the DNA translocation speed. But it still remains a challenge to reduce the translocation speed enough to obtain distinct current signatures corresponding to individual nucleotides.

With the increased understanding of translocation mechanics and the ability to engineer it, solid-state nanopores were used for more innovative applications. Wanunu and Meller [11] reported an extensive study on chemical modification of solid-state pores. They functionalized the pores with self-assembly of six different

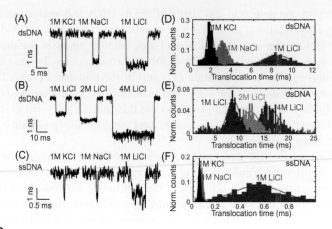

FIGURE 6.3

Effect of LiCl concentration on DNA translocation speed. (A) Comparison of translocation speed and amplitude of current drop for ds-DNA while using 1 M KCl, 1 M NaCl, and 1 M LiCl. (B) Comparison of different concentrations of LiCl; note that higher concentration leads to larger current drop and slower translocation speed. (C) Comparison of three electrolytes for ss-DNA. (D), (E), and (F) represent translocation time histograms corresponding to (A), (B), and (C), respectively.

Source: Adapted with permission from Ref. [10]. Copyright (2012) American Chemical Society.

organo-silane molecules to create "smart" sensors with controlled surface properties. Nanopores have also been functionalized with DNA by Iqbal et al. [12]; specifically with human placental lactogen (HPL)-DNA (using organo-silanes) for selective translocations of ss-DNA samples. When perfectly complementary (pc) and mismatched (mm) DNA structures were introduced in the *cis* chamber, a remarkable difference in translocation time and event frequency was observed. The event frequency of pc-DNA was about 30 times the frequency of mm-DNA (for a single base mismatch).

Solid-state nanopores were also shown to detect single nucleotide polymorphism (SNP) using the DNA—protein complex dissociation kinetics [13]. SNP is a DNA sequence variation that occurs when a single nucleotide differs between two alleles of the same gene. Zhao et al. [13] used nanopores with diameters 2−3.5 nm to translocate restriction enzyme EcoR1—DNA complexes and observed translocation events >10 s. These long events were the result of trapping of the complexes in the pore. When the voltage bias was increased, the electric field induced a shear force between DNA and the enzyme and resulted in dissociation of the complex and subsequent translocation of DNA to the *trans* chamber. Using a $2.7 \times 3.4 \pm 0.2$ nm pore, the threshold voltage for disrupting the DNA—enzyme (EcoR1 bound to its cognate site on the DNA) bond was $V = 1.75$ V; however, a single base mutation G→T on the first cognate site from $5'$ end reduced the bond dissociation threshold to $V = 0.5$ V. Since SNPs can predict human susceptibility to diseases and the response to pathogens, chemical, drugs, and vaccines, this solid-state nanopore-based detection strategy can be developed into a clinical diagnostic platform.

Besides DNA—protein complex dissociation kinetics, translocation of protein coated DNA has also been investigated using solid-state nanopores [14—16]. Dekker's group immobilized recombination protein A (RecA) onto ds-DNA structures and studied these DNA—protein complexes using 20—35 nm pores. Concept of translocation of protein coated DNA is depicted in Figure 6.4A. The electrical conductance blockade for DNA coated with RecA was observed to be 12 times the current blockade due to bare DNA when 1 M KCl was used (Figure 6.4C). RecA was chosen for these studies because it is an important protein essential for repair and maintenance of the DNA. This protein is also central to homologous recombination in prokaryotes. RecA associates with ss-DNA and catalysis pairing between ss- and ds-DNA thus facilitating exchange of the strands. The ability to detect localization of proteins like RecA (or other proteins or transcription factors) along the DNA strand can be useful in genomic/proteomic screening applications in the future.

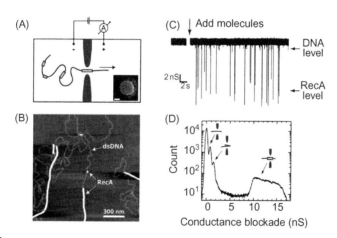

FIGURE 6.4

Experimental details for DNA—protein complex translocation. (A) Schematic diagram for the experiment depicting DNA (purple) coated with RecA protein (orange) translocating through the pore, insert shows TEM image of a 30 nm pore used for experiments. (B) AFM image of the analyte, RecA protein is shown in white. (C) Translocation events observed by the authors. (D) Current drop histogram shows multiple peaks corresponding to different conformations of bare DNA and the protein coated regions in DNA. (For interpretation of the references to color in this figure legend, the reader is referred to the web version of this book.)

Source: Adapted with permission from Ref. [15]. Copyright (2009) American Chemical Society.

Solid-state nanopores have also been used to investigate peptide nucleic acid (PNA) tagged ds-DNA molecules [17]. PNAs are a class of nondegradable oligonucleotide mimics consisting of a synthetic polypeptides backbone onto which nucleobases are attached as side chains. PNAs are used as DNA probes and have found applications in molecular biology procedures, diagnostic assays (detection, quantification of DNA mutations), and antisense therapies. In their study, Singer et al. [17] hybridized two 3.5 kbp fragments (F1 and F2) with PNA probes. PNA probes had eight-base sequence specificity only for the second fragment (F2) and were expected to hybridize with it, whereas fragment F1 was the internal control. When the PNA tagged DNA fragments were translocated through a 4 nm pore, PNA binding to F2 could be clearly distinguished using the current signatures. As demonstrated by Singer et al., the use of nanopore system for lengthwise detection of PNA hybridization along the ds-DNA can emerge as a very valuable diagnostic platform.

Besides DNA analysis, nanopores have also been used for the analysis of RNA molecules [18,19]. Skinner et al. performed a systematic study in which they translocated DNA, ds-RNA, and RNA homopolymers poly(A), poly(U), and poly(C) through 10 nm diameter pores drilled in 20 nm thick freestanding SiN membranes. They performed these experiments at voltages ranging from 100 to 600 mV and calculated mean conductance change $\Delta G = \Delta I / V$ (ΔI is the current drop corresponding to an event and is obtained as $\Delta I = I_{\text{open\_pore}} - I_{\text{blocked}}$). When ΔG was plotted against different voltage bias (Figure 6.5) for all samples, a very interesting phenomenon was observed. As the voltage bias was increased, double-stranded molecules were observed to block more conductance than single-stranded molecules [18]. According to Skinner et al., flexible ss-molecules get stretched under the influence of high electric field close to the pore and thus block less conductance, whereas the ds-molecules are relatively less flexible and therefore block more conductance at higher voltages. When a mixture of ds-DNA and poly(A) was introduced in the *cis* chamber, this differential behavior of ss- and ds-molecules at high voltages could be used to distinguish the two molecules.

The same group has also reported a force measurement study on ds-RNA in the nanopore [19]. For this study, ds-RNA molecules were immobilized on 1.87 ± 0.03 μm polystyrene beads and these beads were optically trapped near the nanopore using an optical tweezers setup. When the voltage bias was applied, negatively charged RNA was pulled into the pore and it was accompanied by proportional translation shift in the position of the microbead. Current blockade and microbead displacement were used to calculate the force per voltage for ds-RNA and it was compared with ds-DNA data. The force per voltage for ds-RNA was reported to be slightly lower than that for ds-DNA. When using a 22 nm pore, the reported values were $f_{\text{dsRNA}} = 0.11 \pm 0.02$ pN/mV and $f_{\text{DNA}} = 0.14 \pm 0.03$ pN/mV. The significance of such force measurement experiments lies in unraveling and studying the secondary structure of ss-molecules.

Discrimination between DNA, RNA, and tRNA; and the detection of miRNA was achieved using thin SiN nanopores by Wanunu et al. [20]. In order to

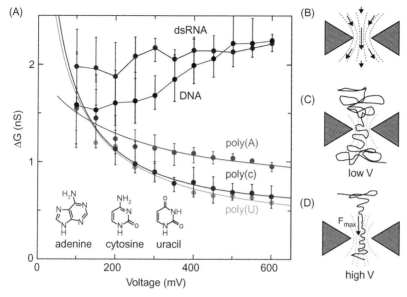

FIGURE 6.5

Voltage-dependent translocation dynamics for ds- and ss-polynucleotides. (A) Variation in ΔG with applied voltage for ds- and ss-polynucleotides. See text for details. (B) Force experienced by charged particles in inhomogeneous electric field inside the pore depends on field strength at particle's location. (C) and (D) Represent the stretching behavior of homopolymers inside the pores. High voltage leads to stretching of flexible ss-molecules and as a result it occupies less volume in the pore and block less conductance.

Source: Adapted with permission from Ref. [18]. Copyright (2009) American Chemical Society.

improve the SNR for nanopores drilled in SiN membranes, 60 nm thick SiN membranes were thinned down to 6 nm. These membranes were etched using SF_6 (etch rate of 1 nm/s) to reduce the thickness and then 4 nm pores were drilled in the thinned membranes using regular procedures. When a 3 kbp ds-DNA was translocated through a 4 nm pore (drilled in membranes of different thickness) at 21°C and 300 mV, a remarkable increase in signal amplitude was observed for pores in thinner membranes. These pores were then used to discriminate the translocation of DNA from RNA and tRNA, and three distinct peaks ($\Delta I_{DNA} = 0.54$ nA, $\Delta I_{RNA} = 0.92$ nA, and $\Delta I_{tRNA} = 1.8$ nA) corresponding to three molecules were obtained in current histogram as shown in Figure 6.6.

The same setup was further used to demonstrate detection of liver-specific miRNA miR122a. To be able to detect specific miRNA, it was first enriched from cellular RNA using sequence specific binding to RNA probe. The probe: miRNA duplex was separated from other cellular RNA using p19-protein functionalized magnetic beads (p19 binds only to ds-RNA in size-dependent manner and does not bind to ss-RNA, tRNA, or rRNA). The presence of miRNA in enriched

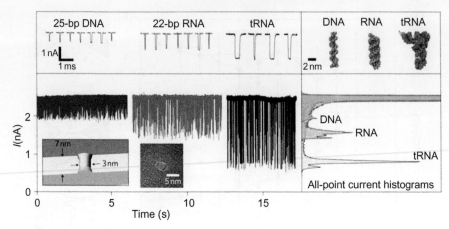

FIGURE 6.6

Discrimination of small nucleic acid using nanopores. 25-bp DNA, 22-bp RNA, and tRNA were distinguished using 3 nm diameter pore drilled in a 7 nm membrane. Continuous current–time traces for the three molecules are shown in blue, red, and black, respectively. Sample events are shown above continuous traces for corresponding molecules. Right panel shows crystal structure model for three molecules and the all-point current histogram. The three molecules could be distinguished based on their current amplitudes in the current histogram. (For interpretation of the references to color in this figure legend, the reader is referred to the web version of this book.)

Source: Adapted with permission from Ref. [20]. Copyright (2010) Macmillan Publishers Ltd.

probe:miRNA samples was detected using the nanopore system and compared with the current traces obtained using commercial probe:miR122a samples. This work demonstrated the application of nanopore system to electronically sense miRNA using small sample volumes and without the need for labeling or surface immobilization of the probes. The reported nanopore-based method for small RNA detection has the potential to be developed into an economic and efficient RNA sensing platform and may prove to be an attractive alternative to the existing techniques.

In addition to the improvement in SNR reported for thin SiN membranes by Wanunu et al. [20], thin membranes may also improve the resolution of DNA sequence detection because the current drop in the pore will correspond to fewer nucleotides due to reduced length of the pore. This idea has led to the development of graphene-based nanopore sensors. Graphene is a thin two-dimensional sheet of carbon with thickness comparable to the spacing between adjacent nucleotides on a DNA strand. Graphene-based devices consist of nanopores drilled in monolayer or multilayers of graphene sheets supported by SiN membranes. Figure 6.7 presents the conceptual diagram for DNA detection using SiN membranes and thin graphene layers.

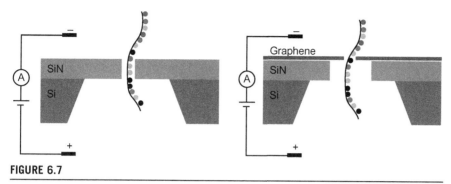

FIGURE 6.7

Schematic for detection using SiN membranes and graphene. Thickness of SiN membranes used for DNA translocation is generally in the range of 10−50 nm and many nucleotides occupy the space in the pore simultaneously. When using graphene, fewer nucleotides occupy the pore at any time which helps to achieve high detection resolution.

The first report on the use of graphene nanopores for DNA analysis was published by the Drndic group at UPenn [21]. They used nanopore devices with sub-10 nm pores drilled in chemically grown 1−5 nm thick (3−15 monolayers) graphene sheets. It was reported that current signals from bare graphene nanopores were noisier than their SiN counterparts and resulted in nonuniform current amplitudes during DNA translocations. The noise was attributed to pinholes in the graphene sheets and to incomplete wetting of the devices due to hydrophobic nature of the chemically grown graphene. As a result, only 10% of the 50 devices they made were in a functional state and could detect DNA translocations. The problems were addressed by atomic layer deposition of TiO_2 which lowered the noise as was shown by power spectral densities (PSD) for bare and TiO_2 coated pores.

Subsequent work on graphene nanopores has been reported by Garaj et al. (Golovchenko Group), Schneider et al. (Cees Dekker Group), and Venkatesan et al. (Bashir Group). Garaj et al. fabricated 5−23 nm diameter pores in freestanding graphene films synthesized by chemical vapor deposition (CVD) [22]. By fitting the measured ionic conductance data to computer calculations, they determined that conductivity of single layer graphene (0.6 nm thick) was linearly proportional to the pore diameter. They demonstrated ds-DNA translocation through a 5 nm pore and detected linear and folded conformations of the molecule. Using computer simulations, they demonstrated that a 0.6 nm thick graphene sheet can achieve a theoretical resolution of 0.35 nm, which suggests that graphene nanopore-based sensors may be able to discern individual nucleotides as the DNA threads through the pore. The limiting factor in achieving this resolution is the DNA translocation speed through the graphene pores which should be brought down to 1 nucleotide/ms from current levels of several 10 s of nucleotides/μs [23].

In contrast to the other groups, Schneider et al. [24] fabricated their sensors by obtaining graphene through mechanical exfoliation from graphite on SiO_2. They advocated an exfoliation method over chemical synthesis as exfoliation results in fewer defects and gives control over the number of layers. The current—voltage characteristics for several pores ranging from 5 to 25 nm were measured and upon fitting the ionic conductance data they concluded that conductance of the pore varied as a function of square of the diameter of the pore as opposed to the linear relationship reported by Garaj et al. Further, detection of the ds-DNA translocation through the graphene nanopores was demonstrated and multiple peaks corresponding to linear and folded conformations of DNA were obtained in the current drop histogram.

Recently, Venkatesan et al. [25] reported the use of stacked graphene—Al_2O_3 nanopores for detection of DNA and DNA—protein complexes. The multilayer architecture fabricated by stacked layers of graphene and Al_2O_3 resulted in less noisy and more robust nanopores. The layering of graphene and the dielectric (Al_2O_3) also allowed for electrical biasing of embedded graphene electrode which can be used for three-terminal control over DNA translocation speed. For device fabrication, CVD grown graphene (g1) was transferred onto preformed Al_2O_3 membrane with 300—350 nm pore (Figure 6.8A and B). Thereafter, metallic Al was evaporated on graphene which was followed by atomic layer deposition of Al_2O_3 (d1). Another graphene layer (g2) was then transferred and Ti/Au contact was formed on the edge of g2 (this contact can serve as the third terminal) and it was followed by a second Al_2O_3 layer (d2) stacking (Figure 6.8C). Once the stack was ready, sub-10 nm pores were drilled using focused electron beam (Figure 6.8D). This nanopore architecture was demonstrated to be able to resolve linear and folded conformations of ds-DNA and detect the presence of RecA

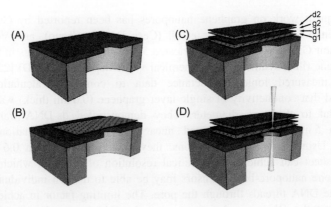

FIGURE 6.8

Fabrication sequence for stacked graphene—Al_2O_3 nanopores. See text for details.

Source: Adapted with permission from Ref. [25]. Copyright (2012) American Chemical Society.

protein on the DNA. This is the first report on protein detection using the graphene nanopores.

Graphene nanopores allow for high detection resolution and if the translocation speed and noise in these nanopores can be reduced, then they may help to realize the goal of nanopore-based DNA sequencing.

6.3 DNA unzipping

DNA unzipping is the process of unwinding the strands of ds-DNA, resulting in two ss-DNA molecules. Unzipping is fundamental to cellular processes like replication and transcription and for *in vitro* techniques like polymerase chain reaction (PCR). The advent of single molecule force spectroscopy techniques like optical tweezers, magnetic tweezers, and atomic force microscopy has made individual DNA molecule accessible for studying this important biophysical phenomena. These techniques allow the study of single molecule level unzipping characteristics of DNA that are generally occluded by ensemble averaging of the information. However, such analytical tools require DNA molecules to be tethered to the bead or the probe for analysis and can achieve only very limited throughput. Solid-state nanopores can be used to achieve similar analytical resolution at much higher throughput, without the need for DNA labeling or immobilization. Using nanopores with diameters smaller than that of ds-DNA but larger than the diameter of ss-DNA ($1.5 < d < 2.5$ nm), a ds-DNA molecule can be sheared and be forced to unzip. Simultaneously, monitoring the force required to unzip the molecule and the progression of the unzipping process can then reveal important biophysical information about ds-polynucleotides.

Zhao et al. [26] used the hairpin DNA (hp-DNA) to investigate the threshold voltage required for translocation of the molecules through a SiN pore and found that it depends on the nanopore diameter and the secondary structure of the DNA. The hp-DNA used by Zhao et al. are the secondary structures formed when a single-stranded polynucleotide folds onto itself allowing complimentary regions to interact via hydrogen bonding resulting in short helical double-stranded regions. Hairpins typically consist of a stem, the region of self-complementary base pairing, and a loop, the single-stranded region between the sections forming the stem, and are regularly found in DNA and RNA molecules. If an hp-DNA molecule is forced through a nanopore with diameter <2.5 nm, it can only translocate if the force induced by the applied voltage is more than the threshold force required to break its secondary structures (or unzip the stem region of the molecule). Zhao et al. investigated different voltage levels and found that threshold voltage required to achieve hp-DNA translocation through a pore with diameter $1.5 < d < 2.3$ nm was $V > 1.5$ V. When smaller pores with diameters between 1.0 and 1.5 nm were used, the threshold voltage reduced to $V < 0.5$ V as the stem unzipped at lower force in the smaller pores.

The hairpins for this study were carefully designed to avoid any unwanted secondary structures while satisfying the stability specifications. Authors used two hairpins: first 150-mer hairpin had a 12-bp long stem with a 76-bp loop and a 50-mer poly-dA overhang. The second hairpin was a 110-mer with a 10-bp stem and 6-bp intervening loop and also had a 50-mer poly-dA overhang. They were prepared by heating the solution to 75°C for 10 min followed by quenching to 4°C. The hairpins were then diluted in 10 mM Tris, 1 M KCl, pH 8.0, and were introduced in the flow cell at a concentration of 0.1 μg/μl. The devices consisted of nanopores of different pore sizes drilled in 10 nm thick SiN membranes. Due to the limitations of their detection apparatus, Zhao et al. [26] used quantitative polymerase chain reaction (qPCR) to determine if the DNA actually went through the pore. For this, the extract from the *trans* chamber was concentrated using a centrifugal filter and the DNA was amplified using qPCR and then analyzed using gel electrophoresis. Gel electrophoresis confirmed the translocation of hp-DNA across the nanopores and the results were used to determine the effect of pore diameter and applied voltage on achieving translocations. Authors suggested that the synthetic nanopores are like molecular gates and can be used to discriminate the DNA secondary structures.

The idea of using the solid-state nanopores to discriminate DNA structures was also explored by McNally et al. [27]. They demonstrated unzipping of the DNA molecules using synthetic sub-2 nm pores and used high-bandwidth electrical measurements for direct readout of the individual unzipping events, obviating the need for PCR amplification used by Zhao et al. [26] They also demonstrated that a single-stranded overhang is required to pull the duplex DNA into a sub-2 nm pore by using three different DNA molecules. Figure 6.9 shows the three samples used by the authors and the corresponding dwell time distributions for these molecules. The first sample was a single-stranded 120-mer poly-d(A) chain; the second sample was the blunt ended ds-DNA consisting of a 24-bp duplex region with a six-base loop. The third sample was a hybrid ss-/ds-molecule with a 100-mer strand hybridized to a fully complementary 50-mer sequence. The hairpin molecules were prepared by heating the oligonucleotides to 95°C followed by sudden cooling on ice, whereas the DNA hybrids were prepared by slow cooling using a thermo cycler. When the three different samples were translocated through the pores, ss-DNA (first sample) resulted in short-lived events with a peak at ~24 μs followed by an exponential decay with $\tau = 8.8 \pm 0.4$ μs. The blunt ended hairpin (second sample) had similar translocation kinetics with peak at 24 μs followed by an exponential decay with $\tau = 21.1 \pm 1.2$ μs. However, the dwell time for the third sample was remarkably different from the other two samples with a peak at ~40 μs followed by a biexponential decay with $t_0 = 40 \pm 6$ μs and $t_1 = 240 \pm 22$ μs.

It was suggested that in the case of the third sample, once the ss-portion of the molecule threaded through the pore, the ds-portion encountered <2 nm pore and it could retract back, translocate through the pore without unzipping or actually unzip due to the shear force experienced by the overhang region. The authors ruled out the first two scenarios based on probability of retraction and the data obtained

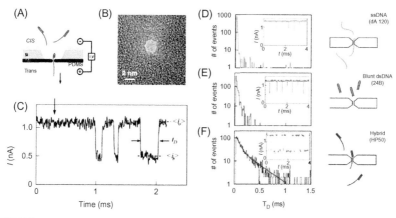

FIGURE 6.9

DNA unzipping using solid-state nanopore. (A) Schematic representation of the experimental setup. (B) TEM image of a 2 nm diameter pore used for unzipping experiments. (C) Current–time trace recorded upon translocation of 24-bp duplex through a 2 nm pore. Arrow indicates addition of the DNA sample to the *cis* chamber and $<i_o>$, $<i_b>$, and t_d are the open current level, blocked current level, and the event duration, respectively. The right panel, (D), (E), and (F) illustrate the three samples used by the authors and the corresponding current traces and event time histograms. See text for details.

Source: Adapted with permission from Ref. [27]. Copyright (2008) American Chemical Society.

from translocation of the hairpin samples. They correlated the kinetic data obtained from the third sample to unzipping phenomena and attributed the characteristic timescales in biexponential decay to collision (t_0, fast timescale) and unzipping (t_1, slow timescale) events. Further, it was demonstrated that this technique could be used to detect base pair mismatch. Two samples with 24-mer duplex region and 50-mer overhang showed remarkably different t_1 timescale when one of the samples had two base mismatches, although they displayed almost identical motilities on nondenaturing poly-acralamide gel electrophoresis (PAGE).

These reports demonstrate the utility of solid-state nanopore-based sensors for discriminating secondary structures of DNA and for high throughput detection of DNA mutation and sequence variability.

6.4 Optical detection

Alongside electrical detection of DNA translocation, optical readout modality has been developed [28–30]. Optical means for detecting translocation events can allow for parallel detection using nanopore arrays and potentially result in high throughput sensing at the single molecule level.

Use of optical detection strategy was first reported by Chansin et al. [28]. They used nonporous membrane for optical detection of fluorescently labeled λ-DNA molecules and demonstrated parallel readout using nanopore arrays. For nanopore fabrication, 200 nm thick SiN membranes were coated with 100 nm thick aluminum layer by thermal evaporation followed by drilling 250–300 nm diameter pore array using focused ion beam (FIB). The nonporous membrane was mounted on a glass slide coated with indium tin oxide (ITO) as shown in Figure 6.10. Glass was coated with ITO to serve as the positive electrode. Moreover, because ITO is transparent, it did not interfere with optical detection. The opaque aluminum coating on the SiN membrane prevented any labeled DNA in the analyte chamber to be illuminated, resulting in low background noise.

DNA molecules were driven across the pore upon applying the electric potential. Excitation light was focused on aluminum membrane using a high numerical aperture objective and the same objective was used to collect fluorescence signals (generated by translocation events) and feed them to the electron multiplying charged couple device (EM-CCD) camera. Using this setup, simultaneous event detection from nanopore array consisting of nine pores could be achieved. The effect of voltage bias on event frequency and translocation speed was also studied and a higher frequency and lower translocation speed was observed with the increasing voltage. This study presented a novel approach in nanopore detection and demonstrated that it is possible to achieve parallel and high-resolution DNA sensing by optical means.

An interesting integration of optical and electrical sensing methods was also reported by Soni et al. [29]. They demonstrated simultaneous and in-sync electrical and optical detection of analytes translocating across the nanopore. Using this integrated system, temporal as well as positional information about the analyte could be obtained which is not possible only with the electrical sensing. The authors used total internal reflection fluorescence (TIRF) excitation to probe molecules in a localized zone closed to refractive index match interface which excited the target

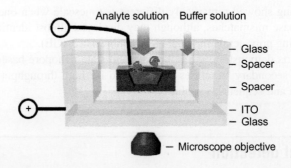

FIGURE 6.10

Setup used for optical detection of DNA translocation by Chansin et al. [28] See text for details.

Source: Adapted with permission from Ref. [28]. Copyright (2007) American Chemical Society.

analyte molecules without producing any background noise. Using TIRF also enabled the use of wide-field optics which can be used to achieve high throughput sensing by simultaneous probing of many pores. The similar system was also used by this research group to develop a DNA sequencing strategy by converting the target DNA strands into a binary code of oligonucleotides and then detecting the converted sequence using optical means [31]. This sequencing strategy is discussed in more detail in the next section.

6.5 **DNA sequencing**

The sequence of bases on a DNA strand contains a wealth of genetic information which is used in building and maintaining an individual. This sequence is unique for every individual and determines the individual's hereditary traits and his/her susceptibility to diseases. This information can be used to tailor the conventional therapeutic approaches and deliver more personalized medicine based on an individual's genetic predisposition of responsiveness to a therapeutic regimen. It is believed that easy availability of a patient's genetic data and the development of personalized medicine can greatly improve treatment efficacy for deadly diseases. Such a therapeutic approach is hindered by the current sequencing techniques which are very inefficient, slow, and outrageously expensive. These obstacles can be overcome by the use of nanopore sensors which may sequence DNA in a more efficient and economic manner. As discussed in earlier sections, nanopores have been shown to detect ss- and ds-DNA, its secondary structure, base mismatches, and SNPs by direct electrical readout, thus demonstrating the potential and promise of this new tool for next generation DNA sequencing.

Over the past decade, many efforts have been made to enhance the sensing capability of nanopore devices to be able to use them for DNA sequencing, but the goal has so far not been fully realized. The efforts made in the direction of using solid-state nanopore devices for DNA sequencing can be clustered into four categories.

6.5.1 **Direct ionic current measurement**

It involves driving the DNA molecule across the nanopore and using current signatures to discern the sequence of nucleotides along its strand. This approach is institutive and was the key idea behind the development of nanopore-based sensors; however, experimentally it lacks spatial and temporal resolution required to distinguish individual nucleotides. The basic requirements that direct ionic current measurement-based method needs to fulfill in order to be used for DNA sequencing are:

a. *Spatial resolution*: Conventional sensors have nanopores drilled in 10−50 nm thick SiN membranes. When DNA is driven through the pore, at any time, several nucleotides are present inside the pore and contribute to the current drop. Such a situation makes it impossible to distinguish between adjacent nucleotides on the DNA strand. This challenge has been addressed by using thin SiN membranes [20] and by the use of graphene nanopores [21,22,24,25] as discussed previously. Garaj et al. [22] used computer simulations to show that nanopores drilled in 0.6 nm thick graphene sheets can achieve single nucleotide resolution and discern subtle features of a DNA molecule.

b. *Temporal resolution*: As discussed earlier, the translocation speeds of DNA through solid-state nanopores is in the range of $10-100$ nucleotides/μs. The measurement of pico-ampere current levels at such high translocation speeds is beyond the limit of currently used electrical detection apparatus and needs the detector bandwidth to be in MHz region. Efforts have been made to slow the translocation speed [8−10] but in order to achieve reliable detection of individual nucleotides, time spent by each base inside the pore needs to be brought down to ≥ 1 ms [23,32].

c. *Noise in the sensors*: Noisy sensors result in low SNR and make it less probable to be able to detect subtle structural features of the translocating molecule. TiO_2 and Al_2O_3 coatings have been used to reduce inherent 1/f noise of graphene nanopores [21,25].

To be able to use direct ionic conductance measurement to discern individual nucleotides, a wholesome technique needs to be developed that can simultaneously address the challenges outlined above.

6.5.2 Hybridization assisted nanopore sequencing (HANS approach)

If a probe (10−50 base long ss-oligonucleotide) can be hybridized with target single-stranded DNA, the DNA molecule will exist in double-stranded form wherever probe hybridizes with it. When such hybridized DNA molecule is driven through the nanopore, the information about binding sites of the probe can be obtained from the temporal profile of the current trace. And by combining this information with *a priori* knowledge of probe sequence, it may be possible to determine the partial sequence of the target DNA. If the same process is repeated with a library of probes and in massively parallel fashion, one should be able to determine complete sequence of the DNA. This is the principle behind hybridization assisted nanopore sequencing (HANS) or the HANS approach.

The feasibility of HANS approach was demonstrated by Balaguruswamy et al. [33]. With the objective of resolving a few base pair long hybridized structures along the temporal profile of ss-DNA, the authors hybridized three ss-oligomers with partially overlapping sequence to generate trimeric structures, ss-ds-ss-ds-ss, as shown in Figure 6.11A. A polystyrene bead was attached to one of the oligonucleotides to reduce the translocation speed of the complex. When these structures were driven through 10 and 15 nm pores, sequentially arranged two 12-base pair

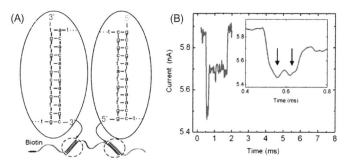

FIGURE 6.11

Detection of DNA hybridization using solid-state nanopores. (A) Trimeric structure designed by the authors consisting of ss-ds-ss-ds-ss segments. (B) Current–time trace obtained when trimeric structures were driven across the pore. The inset shows a magnified view of a representative trimeric complex translocation event. Black arrows indicate the location of ds segments in the complex.

Source: Adapted with permission from Ref. [33]. Copyright (2010) IOP Publishing.

hybridization segments separated by 140-base ss-sequence could be clearly detected.

The results obtained from this study are encouraging as they demonstrate the feasibility of HANS approach. But this is only proof of concept and in order to use this approach for routine DNA sequencing, this concept needs to be further developed. The major challenges for this approach are to accurately determine the locations of probe hybridization, scale up the process to obtain hybridization results for a library of probes, and develop computational routines to construct the DNA sequence from the translocation data.

6.5.3 Optical readout using converted DNA

This concept was developed by Meller's group at Boston University [31] and is based on converting the four-base (A, T, G, and C) sequence of DNA into a binary color scheme followed by an optical readout. As shown in Figure 6.12; first, each base in the original DNA strand is converted into a sequence of two 12-mer oligonucleotides. The sequence information contained in the original DNA is encoded into the sequence of attachment of the two oligonucleotides. For example, if the oligonucleotides are represented by 0 and 1, then adenine (A) can be coded as "11," guanine (G) as "10," thymine (T) as "01," and cytosine (C) as "00." This way, any 10-base sequence AGATTCGATA in the original DNA can be converted into a binary code which may look like 11-10-11-01-01-00-10-11-01-11. Once the DNA is converted, it is hybridized with two types of molecular beacons. These molecular beacons are oligonucleotides complementary to the two oligonucleotides used in making the converted DNA strand and are labeled with two different fluorophores and the quencher. In solution, both the free beacons

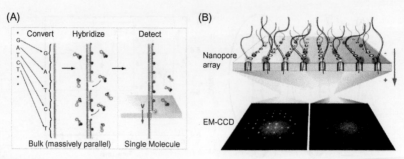

FIGURE 6.12

DNA sequencing by optical detection. (A) Schematic of sequencing strategy proposed by the authors. The target DNA strand is first converted to a sequence of oligonucleotides which is followed by hybridization with fluorescently tagged probes. When the hybridized DNA is driven through sub-2 nm pore, probes are sequentially stripped off generating fluorescent signals. See text for details. (B) Schematic illustration of parallel detection scheme for converted DNA using nanopore arrays.

Source: Adapted with permission from Ref. [31]. Copyright (2010) American Chemical Society.

and ones bound to the converted DNA produce very low background fluorescence because of self-quenching and due to their proximity to the other beacons. After hybridization, DNA is driven through 1.7−2 nm pores which result in stripping off (unzipping) of the hybridized molecular beacons one by one. The stripping process causes un-quenching of fluorescence and results in short-lived photonic bursts which are captured by EM-CCD camera. If the oligonucleotides representing 0 and 1 in the above example are hybridized with the probes labeled with "red (R)" and "yellow (Y)" fluorophores, our 10-base sequence on the original DNA strand may correspond to YYYRYYRYRYRRYRYYRYYY color sequence. This way by recording the color sequence from the hybridized DNA translocating through the pores, the sequence of bases on the original DNA strand can be inferred. As a proof of concept, McNally et al. demonstrated the use of two color beacons as described above and identified individual nucleotides on the target DNA with a high degree of accuracy. They also demonstrated simultaneous optical detection from three nanopores highlighting the potential of this modality for parallel high throughput DNA sequencing.

This optical approach has the following advantages which set it apart from other nanopore-based sequencing methods:

a. As the four-base sequence is reduced to binary sequence of oligonucleotides, it significantly reduces the complexity of the problem.
b. Because the threading of the DNA across the pore involves unzipping of fluorescent tagged complementary oligonucleotide, lower translocation speed can be achieved.
c. The temporal resolution of detection is determined by the time it takes to unzip the fluorescent tagged probes which in turn depends on the length of the

oligonucleotides in the converted DNA. By using longer oligonucleotides, one may be able to further reduce translocation speed and lower detection resolution, though it will slow the overall sequencing process.

d. Optical readout enables massively parallel detection. As one nanopore can probe only one DNA molecule at a time, there is a need for scaling up the detection process (by using nanopore arrays) to achieve high throughput. Unfortunately, ionic conductance-based detection methods cannot be used for simultaneous detection using nanopore arrays; however, such parallel detection is feasible using the optical readout.

This concept is very attractive, but it also needs to overcome a few challenges to be practically used for sequencing. The DNA conversion adds an extra chemical step to the sequencing process and is slow and error prone. Also, the unzipping process makes the process very slow and impractical for whole genome sequencing. Further, it is challenging to fabricate sub-2 nm pore arrays to implement parallel detection strategy.

6.5.4 Transverse electron tunneling

This modality of nanopore sequencing uses quantum mechanical phenomena of electron tunneling to identify individual nucleotides on the target DNA strand. When two electrodes are separated by a distance of a few nanometers and an electric field is applied across the electrodes, the energy barrier is overcome by the electrons resulting in a current (electron tunneling) that varies exponentially with the applied voltage and the separation between electrodes (nanogap). This current is very sensitive to the size, shape, and orientation of molecules existing in the nanogap. It was proposed that if such tunneling probes are integrated into the nanopore device, then the tunneling current through the nucleotides driven across the nanopore can be used to identify the individual bases [34,35]. This approach overcomes the spatial resolution limitation of the nanopore because the nucleotides are not identified by the current drop due to base occupancy in the pore; instead, the transverse tunneling current from sub-nm diameter tunneling probe tip is used to identify the nucleotides. The idea is illustrated by the schematic shown in Figure 6.13. Moreover, the tunneling current decays very rapidly with the distance, and it results in a signal specific to the base in the nanogap with almost no interference from the neighboring bases. This leads to high spatial resolution and molecular specificity [36].

Identification of individual DNA bases using electron tunneling was first reported by Tsutsui et al. [37] from Osaka University. They used nanofabricated mechanically controllable break junctions to realize nanoelectrodes with gap spacing of the order of size of the nucleotides. When nucleotides dissolved in Milli-Q water at a concentration of ~5 μM were introduced in the nanogap and a voltage bias of 0.75 V was applied, current spikes corresponding to the nucleotides were observed. Three of the four nucleotides—GMP, CMP, and TMP—could be

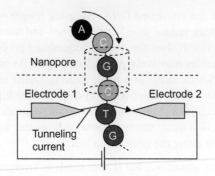

FIGURE 6.13

Conceptual illustration of integration between nanopore sensor and tunneling current probe. Nanopore can be used to control the translocation speed and unravel DNA secondary structures; and tunneling current can be used to identify individual nucleotides.

Source: Reprinted with permission from Ref. [36]. Copyright (2011) American Chemical Society.

detected using their setup, but AMP could not be detected due to significant background noise. The authors reported that the magnitude of current spikes was dependent on voltage bias and higher spikes were observed at higher voltages. Also, the current drop value was different for the three nucleotides and could be used to distinguish them from one another.

Because the tunneling current is very sensitive to molecular orientation inside the nanogap, if the same nucleotide is present in the gap in different orientations, it can result in a large variation in the tunneling current. This issue was addressed by Huang et al. [38] by using a chemical linker to orient nucleotides in the same orientation each time, resulting in more reproducible tunneling current. To achieve this, gold electrodes were functionalized with 4-mercaptobenzamide which bound to the gold surface with one end using gold−thiol linkage and had nitrogen (hydrogen bond donor) and carbonyl (hydrogen bond acceptor) groups on the other end which could form hydrogen bonds with the nucleotides in the nanogap. Formation of hydrogen bonds ensured nucleotide existed in a specific configuration inside the nanogap and prevented any free rotation resulting in high SNR. Using these functionalized electrodes, Huang et al. were able to detect and discern deoxyadenosine 5′-monophosphate (dAMP), deoxyguanosine 5′-monophosphate (dGMP), deoxycytidine 5′-monophosphate (dCMP), and 5-methyl-deoxycytidine 5′-monophosphate (d^mCMP) unambiguously. Thymidine 5′-monophosphate (dTMP) could not be detected and the inability to detect was attributed to strong binding of TMP to the gold surface.

Recently, integration of tunneling current probes and nanopores has also been realized [36,39]. The idea behind this integration is that the nanopore can be used to control the translocation speed and unravel the secondary structures of DNA, while individual nucleotides can be detected using the tunneling current. The integration of these two modalities presents a more practical approach to DNA sequencing but until recently it had been hindered by the technical limitation in precisely aligning the position of nanopore and sub-nm scale electrode nanogap. The integration of tunnel junction with the nanopore was first reported by Ivanov et al. [36]. The devices were fabricated as illustrated in Figure 6.14; briefly, 70 nm thick SiN layer was deposited on either side of the wafer followed by fabrication of gold electrodes on the front face with an electrode gap of 2 µm. Following electrode fabrication, another 300 nm thick layer of SiN was deposited over the electrodes which helped in minimizing membrane capacitance and improved its mechanical strength. Further, 5 µm × 5 µm electrode-centered window was opened in SiN on the front face to expose the electrodes and a bigger window was opened in the back face for silicon etching and producing freestanding SiN membrane. Then 50–70 nm diameter pore was drilled in the middle of the gold electrodes using FIB, which was followed by fabrication of platinum nanoelectrodes using electron beam-induced deposition (EBID). As a proof of concept, the authors used this device to demonstrate detection of λ-DNA simultaneously with both ionic and tunneling currents.

In a more recent publication, Tsutsui et al. [39] reported fabrication of electrode embedded nanopores for DNA sequencing using electron tunneling. They developed a self-aligning technique to integrate sub-nm sized electrode gap with 15 nm diameter SiO_2 nanopores. To fabricate the aligned nanoelectrodes, they used electromigration/self-breaking technique which resulted in sub-nm electrode

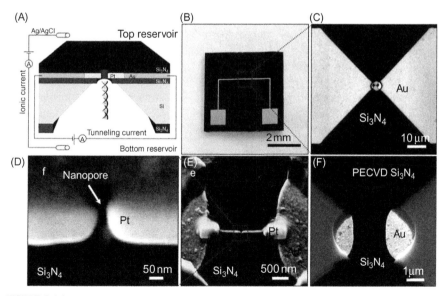

FIGURE 6.14

Fabrication of aligned nanopore and tunneling electrodes. See text for details.

Source: Reprinted with permission from Ref. [36]. Copyright (2011) American Chemical Society.

gap. Using these devices, the authors demonstrated detection and counting of nucleotide sized metal coated fullerenes in liquid. They also demonstrated detection of 6-mer oligonucleotides with TTTGGG sequence and obtained a bimodal distribution corresponding to the two nucleotides.

The tunneling junction integrated nanopores can potentially be used for fast and label-free DNA sequencing. But in order to be practically used for the purpose of sequencing, this technique still needs to overcome the issues of fast translocation speed and tunneling current variations due to thermal fluctuations of molecules in the nanogap.

6.6 Protein analysis with solid-state nanopores

Although DNA sequencing is the major driving force behind the development of nanopore sensors, this versatile single molecule sensing platform has also been used to elucidate size, shape, charge, folding state, binding kinetics, and other biophysical properties of proteins and polypeptides. As compared to DNA, proteins are a more complex analyte for analysis using nanopores. A DNA molecule is generally linear, chemically stable, and has uniform negative charge over a range of pHs. By applying voltage bias across the nanopore, DNA can be easily translocated through it, generally producing clear spikes with quantized current levels. On the other hand, proteins are more complex and exhibit diverse conformational states. They consist of hydrophobic and hydrophilic residues and show different structures and functions based on interaction with the solvent. Moreover, proteins have a heterogeneous charge distribution and may get polarized under the influence of electric field as the positively and negatively charged amino acids are pulled in opposite directions [40].

Despite the complexity of protein structure, nanopore sensing is an attractive platform for protein analysis as it can be used to accomplish single molecule level detection of label-free proteins at their physiological concentrations and in their native environments. Here we discuss some key experiments in the area of protein analysis using solid-state nanopores.

The first instance of using solid-state nanopores for protein analysis was reported by Han et al. [41]. They demonstrated translocation of 66 KDa bovine serum albumin (BSA) molecules using 55 nm diameter pores drilled in 20 nm thick SiN membranes. The experiments were performed with 1 M KCl, 10 mM Tris pH 8, and at this pH BSA had a net negative charge (isoelectric point, pI of BSA is 5.1−5.5). When a potential bias of 100 mV was applied to the *trans* chamber, current spikes with magnitude up to −400 pA were detected. When a potential of −100 mV was applied, a few small spikes were observed and were attributed to bumping events. The authors suggested that the recorded spikes were due to BSA translocations across the pore based on the circumstantial observation but did not verify the claim using any biochemical methods because the concentration of translocated BSA molecules in the *trans* chamber was very low.

The doubts in Han's report about translocation of protein molecules were settled by a subsequent publication from Fologea et al. [42]. For studying BSA translocation, they used 16 nm diameter pores drilled in 10 nm thick SiN membranes. To demonstrate the effect of pH on the charge of BSA, experiments were carried out at both acidic and basic pH (with respect to pI of BSA) using 0.4 M KCl. At pH 7, the protein was negatively charged and applying a bias of 120 mV resulted in spikes with average current drop and event duration of ∼50 pA and ∼110 μs, respectively; whereas, at pH 4.3 the protein was positively charged and translocated only when a voltage bias of −120 mV was applied which resulted in an average current drop and event duration of ∼71 pA and ∼269 μs. To verify that the spikes were indeed due to translocation events, an experiment was carried out for ∼50 h in order to enrich the number of BSA molecules in the *trans* chamber and about 2×10^7 blockade events were registered during this time. The solution from the *trans* chamber was then collected and BSA was detected using an enzyme-immunometric kit. Results indicated a concentration of ∼1.8 pg BSA which corresponded to about 10^7 molecules. Further, to illustrate how the current signatures depend on size and structure of the protein molecules, similar experiments were performed with fibrinogen (340 KDa protein with similar charge as BSA). The current drop and event duration values obtained for fibrinogen were remarkably different from those of BSA and they could be used to distinguish between the two proteins. The authors suggested that the current drop, event duration, and integrated area of blockade can be used to infer the properties of an unknown protein by correlating these parameters for a marker protein of known charge, size, and conformation.

In a later publication, Han et al. [43] demonstrated label-free detection of four different proteins and protein−protein complexes using solid-state nanopores. For this study, they used nanopores smaller (28 nm diameter) than the ones used in their earlier report. They reported label-free detection of BSA, ovalbumin (OA), avidin (AV), and streptavidin (SAV) over a range of pHs and quantified the event frequency at each pH. The maximum event frequency was generally detected around the pI of the proteins. Furthermore, the authors detected protein−protein interactions and demonstrated the use of solid-state nanopores for label-free, amplification-free, and immobilization-free immuno-assay development. This molecular interaction was demonstrated using β-human chorionic gonadotropin (β-hCG), a pregnancy marker abundantly available in body fluids of pregnant females, and anti-β-hCG monoclonal antibodies. When anti-β-hCG antibodies were added to the *cis* chamber, a few current spikes corresponding to the antibody were observed. Following antibody, OA was added as a noninteracting control molecule and this resulted in two levels of current spikes corresponding to antibodies and OA. Finally, the β-hCG antigen was added and it resulted in dose-dependent disappearance of antibody spikes, contrary to what was expected. The disappearance of antibody spikes was attributed to change in charge or mobility of antibody by complex formation with the antigen. Since nanopores can detect proteins at extremely low concentrations, they may develop into an effective tool for protein detection and disease diagnosis, but it requires a more systematic study on the behavior of protein–protein complexes inside nanopore.

In the earlier studies, protein translocations exhibited some intriguing and unexpected behavior based on pH and voltage bias and it needed further exploration. Firnkes et al. [44] suggested that translocation of proteins across the SiN membranes was not straightforward; rather, it was a synergistic action of diffusive, electrophoretic and electroosmotic forces. The authors used AV as the model protein and measured zeta potential (ζ) of the protein and the SiN membrane as a function of the pH. They suggested that one could predict the translocation direction of the protein based on the difference between its zeta potential and that of SiN membrane as this difference ($\zeta_{protein} - \zeta_{pore}$) directly translates into the balance of electrophoretic and electroosmotic forces. Using this model, translocation of AV was demonstrated at different pHs. As can be seen in Figure 6.15, different translocation direction and variable event frequency were observed for experiments at different pH depending on the relative magnitudes of electrophoretic and electroosmotic forces. When these two forces canceled each other out, translocation of protein was dominated by diffusive transfer and it occurred at a significant rate.

Another interesting work on the behavior of proteins inside solid-state nanopores was reported by Niedzwiecki et al. [45]. They explored nonspecific adsorption of BSA on SiN surface and its effects on protein translocation. The events were classified as the short-lived events resulting from protein translocation without interaction with the nanopore and the long-lived events which resulted from protein adsorption in the lumen of the pore. The phenomena were explored using many different pores varying in diameter from 3 to 25 nm, all drilled in 20 nm thick SiN membranes. The short-lived events generally increased with increasing protein concentration and voltage bias but not the long-lived ones. For the longer events, it was hypothesized that BSA adsorbs in two different orientations. One of these orientations resulted only in decrease of the baseline current but the other orientation lead to significant current fluctuations perhaps because of vibration of the adsorbed protein.

Besides protein translocation and protein–protein complex detection, solid-state pores have been employed to study protein folding and to distinguish between different folding states. The folding of a linear amino acid sequence into a three-dimensional protein is a crucial process in biology that yields a particular functional form. Due to the vast number of degrees of freedom of the polypeptide chain and the inability to sample all possible conformations in a short amount of time (i.e., μs), folding is currently understood to occur on a complex energy landscape which directs the protein to the singular native state (i.e., the energy minimum). A comprehensive understanding of the process has yet to be achieved and continues to be an area of aggressive research. Once fully understood, the hope is to be able to mitigate or prevent protein misfolding diseases and design proteins with novel functions using protein engineering.

The ability to discriminate between folding states was first reported by Talaga and Li [40] in which they used an extremely thin (10−15 nm) SiN membrane.

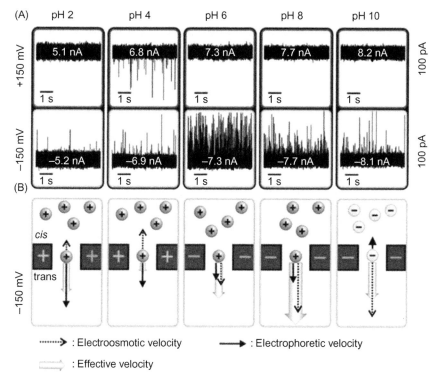

········▶ : Electroosmotic velocity ——▶ : Electrophoretic velocity

⇨ : Effective velocity

FIGURE 6.15

(A) Current traces obtained when +150 and −150 mV were applied to the *trans* chamber for different pH conditions. Current values in white represent the open pore current.
(B) Schematic representation of charges on protein and membrane, and balance of electrophoretic and electroosmotic velocities. Arrow length represents the magnitude of zeta potential.

Source: Adapted with permission from Ref. [44]. Copyright (2010) American Chemical Society.

The membrane thickness, or the effective membrane thickness taking into account access resistance, was a key factor in the analysis of the protein β-lactoglobulin which was chemically unfolded using urea. Since the contour length of the protein is larger than the effective membrane thickness, the possibility for segments of the protein to exist outside the pore was possible. In such a scenario, the region of protein being sensed changes throughout the translocation process and furthermore, the charge within the pore and thus the forces on the protein changes as well. The authors made the hypothesis that this led to the existence of stall points in which the protein momentarily stopped moving within the pore due to the segment of protein within the pore having zero net charge. This study shed light on the importance of membrane thickness in relation to protein translocation kinetics and its role in detecting chemically denatured proteins.

Further analysis of unfolding was demonstrated by Freedman et al. [46], in which BSA was unfolded using thermal, chemical (e.g., urea), and electric field methods. In this work, the membrane thickness was larger (50 nm) and led to new observations, specifically when using urea. As opposed to Talaga and Li's observation that the blocked volume due to the unfolded protein is actually reduced with the addition of urea due to segments of the protein leaving the pore, the larger membranes yielded an increase of the excluded volume due to the less compact state of the protein, increased exposed surface area, and the fact that the majority of the protein was likely to be residing inside the pore. More interestingly, the authors reported that the effect of urea was nearly undetectable at higher applied voltages, leading to the hypothesis that the electric field was the dominant force influencing protein folding. Furthermore, the dominant peak had an excluded volume smaller than that of the native and unfolded states at lower voltages. Since the excluded volume equation is a function of molecular shape, this suggested that the electric field unfolded (i.e., changed the shape or molecular structure) of both the native protein and the chemically destabilized protein. This demonstrated that electric fields are powerful denaturants capable of overcoming the molecular forces holding a protein together and even manipulating the unfolded protein's structure.

The discovery that a solid-state nanopore can locally unfold a single protein using an electric field is extremely useful and holds the potential to probe a number of biophysical properties such as stability and binding kinetics. The unfolding occurs due to the fact that proteins have a heterogeneous charge distribution along its amino acid chain. In an electric field, the charged residues get pulled along the axis of the field and this leads to structural instability and eventual unfolding. This represents a new mechanism of nonequilibrium single molecule unfolding which differs from other methods such as atomic force microscopy and optical tweezers. As such, new protein structures can be probed and new regions of the conformational space can be analyzed to obtain a more comprehensive look into protein folding.

Further, the performance of synthetic nanopores for single molecule analysis can be improved if one can control the surface chemistry of the pore and prevent nonspecific protein−nanopore interaction. This has been demonstrated by modifying the nanopore surface with bio-functional molecules which resulted in better performing solid-state pores. Yusko et al. [47] modified the nanopore with fluid lipid bilayer to enhance its performance for protein sensing. The coating of the pore with lipid bilayer conferred sub-nanometer control on pore diameter and the incorporation of mobile ligands in the bilayer resulted in highly specific surfaces which increased the translocation time and prevented pore clogging. The inspiration for the work came from an insect's natural mechanism of detecting pheromones in the environment; insects achieve this feat by translocating the odorant molecules through lipid coated nanopores present in olfactory sensilla of their antennae. Lipid coatings in these sensory receptors are thought to impart specificity and achieve preconcentration which helps insects to sense and respond to extremely low concentrations of odorant molecules in their surroundings.

To fabricate the nanopores with lipid bilayer, pores in SiN membranes were coated with an aqueous solution of unilamellar liposomes. The thickness of the bilayer and its surface characteristic could be controlled by varying the length of hydrocarbon chain in the lipids and polar head group, respectively. Yusko et al. [47] demonstrated that when SAV was driven through pores coated with biotinylated lipids, affinity-dependent preconcentration and reduction in translocation speed could be achieved. It remarkably lowered the bulk concentration of analyte required to register a significant number of events. When pores without biotin functionalized lipids were used, a 500-fold lower event frequency was reported (Figure 6.16).

Through this work the authors have demonstrated that by using such hybrid pores, many problems associated with solid-state nanopores can be circumvented resulting in more sensitive, specific, durable and sensing devices.

Use of hybrid pores to simulate molecular transport across nuclear pore complex (NPC) was also reported by Kowalczyk et al. [48]. The authors transformed the nanopores into biomimetic NPC by functionalizing the pore surface with phenylalanine–glycine nucleoporins Nup98 or Nup153. To fabricate the hybrid pores, SiN membrane was coated with (aminopropyl)-triethoxysilane (APTES) followed by conjugation with Nup98 or Nup153 using a heterobifunctional cross-linker. The modified pores increased the translocation time for transport receptors Impβ (~2.5 ms) and strongly prevented translocation of other proteins with high specificity. These biomimetic nanopores demonstrated remarkable selectivity for Impβ, which is a defining feature of native NPC.

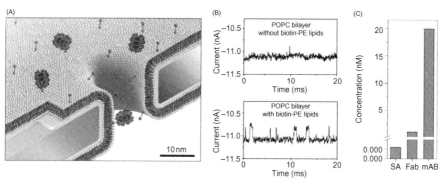

FIGURE 6.16

Lipid bilayer modified solid-state nanopores for protein detection. (A) Conceptual representation of binding of SAV (large red) to biotin-PE (blue small) anchored in the lipid bilayer. This binding leads to selective translocation of SAV across the pore. (B) Current–time traces for SAV detection with POPC bilayers with and without biotin-PE. (C) Minimum bulk concentration of SA, polyclonal anti-biotin Fab, and monoclonal anti-biotin IgG required to detect 30–100 events/s. Biotin-PE: 1-oleoyl-2-(12-biotinyl (aminododecanoyl))-sn-glycero-3-phosphoethanolamine; SA: streptavidin; Fab: antibody fragment; mAb: monoclonal antibody; POPC: 1-palmitoyl-2-oleoyl-sn-glycero-3-phosphocholine. (For interpretation of the references to color in this figure legend, the reader is referred to the web version of this book.)

Recently, Wie et al. [49] reported the use of receptor-modified solid-state nanopores for stochastic sensing of protein molecules. Nanopores were modified by depositing a 40–50 nm layer of gold followed by formation of self-assembled monolayer (SAM) of alkane-thiols which bind to the gold surface by gold–thiol linkage. The pores were receptor-modified by mixing nitrilotriacetic acid (NTA) receptor molecules in the SAM matrix. NTA receptors served as binding sites for His-tagged proteins of interest, and alkane SAM prevented nonspecific sticking of protein to the pore surface. Using these modified pores, reversible binding and unbinding events on single molecule level could be observed in real time. By coupling His-tagged protein A (protein A is specific to Fc region of antibodies) to the NTA receptors, it was demonstrated that these hybrid pores could even differentiate between subclasses of rodent IgG antibodies based on their dwell times.

6.7 Conclusion

Solid-state nanopores have been used to generate a wealth of valuable information for DNA and protein analysis. They have emerged as a powerful tool for studying biophysical properties at single molecule level. We anticipate seeing nanopore-based clinical diagnostics systems and DNA sequencers in the near future. We also hope that as the field continues to grow, these versatile sensors will help to reveal unknown molecular characteristics of various biomolecules.

References

[1] Li J, Stein D, Branton D, McMullan C, Aziz M, Golovchenko J. Ion-beam sculpting at nanometre length scales. Nature 2001;412(6843):166–9.

[2] Li J, Gershow M, Stein D, Brandin E, Golovchenko JA. DNA molecules and configurations in a solid-state nanopore microscope. Nat Mater 2003;2(9):611–5.

[3] Chen P, Gu J, Brandin E, Kim YR, Wang Q, Branton D. Probing single DNA molecule transport using fabricated nanopores. Nano Lett 2004;4(11):2293–8.

[4] Heng JB, Ho C, Kim T, Timp R, Aksimentiev A, Grinkova YV, et al. Sizing DNA using a nanometer-diameter pore. Biophys J 2004;87(4):2905–11.

[5] Storm AJ, Storm C, Chen J, Zandbergen H, Joanny JF, Dekker C. Fast DNA translocation through a solid-state nanopore. Nano Lett 2005;5(7):1193–7.

[6] Storm A, Chen J, Zandbergen H, Dekker C. Translocation of double-strand DNA through a silicon oxide nanopore. Phys Rev E 2005;71:5.

[7] Fologea D, Gershow M, Ledden B, McNabb DS, Golovchenko JA, Li J. Detecting single stranded DNA with a solid state nanopore. Nano Lett 2005;5(10):1905–9.

[8] Fologea D, Uplinger J, Thomas B, McNabb DS, Li J. Slowing DNA translocation in a solid-state nanopore. Nano Lett 2005;5(9):1734–7.

[9] Kim YR, Min J, Lee IH, Kim S, Kim AG, Kim K, et al. Nanopore sensor for fast label-free detection of short double-stranded DNAs. Biosens Bioelectron 2007;22(12):2926–31.

[10] Kowalczyk SW, Wells DB, Aksimentiev A, Dekker C. Slowing down DNA translocation through a nanopore in lithium chloride. Nano Lett 2012.

[11] Wanunu M, Meller A. Chemically modified solid-state nanopores. Nano Lett 2007; 7(6):1580−5.

[12] Iqbal SM, Akin D, Bashir R. Solid-state nanopore channels with DNA selectivity. Nat Nanotechnol 2007;2(4):243−8.

[13] Zhao Q, Sigalov G, Dimitrov V, Dorvel B, Mirsaidov U, Sligar S, et al. Detecting SNPs using a synthetic nanopore. Nano Lett 2007;7(6):1680−5.

[14] Hall AR, van Dorp S, Lemay SG, Dekker C. Electrophoretic force on a protein-coated DNA molecule in a solid-state nanopore. Nano Lett 2009;9(12):4441−5.

[15] Kowalczyk SW, Hall AR, Dekker C. Detection of local protein structures along DNA using solid-state nanopores. Nano Lett 2009;10(1):324−8.

[16] Smeets R, Kowalczyk S, Hall A, Dekker N, Dekker C. Translocation of RecA-coated double-stranded DNA through solid-state nanopores. Nano Lett 2008;9(9): 3089−95.

[17] Singer A, Wanunu M, Morrison W, Kuhn H, Frank-Kamenetskii M, Meller A. Nanopore based sequence specific detection of duplex DNA for genomic profiling. Nano Lett 2010;10(2):738−42.

[18] Skinner GM, van den Hout M, Broekmans O, Dekker C, Dekker NH. Distinguishing single- and double-stranded nucleic acid molecules using solid-state nanopores. Nano Lett 2009;9(8):2953−60.

[19] van den Hout M, Vilfan ID, Hage S, Dekker NH. Direct force measurements on double-stranded RNA in solid-state nanopores. Nano Lett 2010;10(2):701−7.

[20] Wanunu M, Dadosh T, Ray V, Jin J, McReynolds L, Drndić M. Rapid electronic detection of probe-specific microRNAs using thin nanopore sensors. Nat Nanotechnol 2010;5(11):807−14.

[21] Merchant CA, Healy K, Wanunu M, Ray V, Peterman N, Bartel J, et al. DNA translocation through graphene nanopores. Nano Lett 2010.

[22] Garaj S, Hubbard W, Reina A, Kong J, Branton D, Golovchenko J. Graphene as a subnanometre trans-electrode membrane. Nature 2010;467(7312):190−3.

[23] Venkatesan BM, Bashir R. Nanopore sensors for nucleic acid analysis. Nat Nanotechnol 2011;6(10):615−24.

[24] Schneider GF, Kowalczyk SW, Calado VE, Pandraud G, Zandbergen HW, Vandersypen LMK, et al. DNA translocation through graphene nanopores. Nano Lett 2010;10(8):3163−7.

[25] Venkatesan BM, Estrada D, Banerjee S, Jin X, Dorgan VE, Bae M-H, et al. Stacked graphene−Al_2O_3 nanopore sensors for sensitive detection of DNA and DNA−protein complexes. ACS Nano 2012.

[26] Zhao Q, Comer J, Dimitrov V, Yemenicioglu S, Aksimentiev A, Timp G. Stretching and unzipping nucleic acid hairpins using a synthetic nanopore. Nucleic Acids Res 2008;36(5):1532−41.

[27] McNally B, Wanunu M, Meller A. Electromechanical unzipping of individual DNA molecules using synthetic sub-2 nm pores. Nano Lett 2008;8(10): 3418−22.

[28] Chansin GAT, Mulero R, Hong J, Kim MJ, Andrew A, deMello J, et al. Single-molecule spectroscopy using nanoporous membranes. Nano Lett 2007;7(9):2901−6.

[29] Soni GV, Singer A, Yu Z, Sun Y, McNally B, Meller A. Synchronous optical and electrical detection of biomolecules traversing through solid-state nanopores. Rev Sci Instrum 2010;81:014301.

[30] Hong J, Lee Y, Chansin GAT, Edel JB, Demello AJ. Design of a solid-state nanopore-based platform for single-molecule spectroscopy. Nanotechnology 2008;19: 165205.

[31] McNally B, Singer A, Yu Z, Sun Y, Weng Z, Meller A. Optical recognition of converted DNA nucleotides for single-molecule DNA sequencing using nanopore arrays. Nano Lett 2010;10(6):2237−44.

[32] Deamer DW, Branton D. Characterization of nucleic acids by nanopore analysis. Acc Chem Res 2002;35(10):817−25.

[33] Balagurusamy VS, Weinger P, Ling XS. Detection of DNA hybridizations using solid-state nanopores. Nanotechnology 2010;21(33):335102.

[34] Zwolak M, Di Ventra M. Electronic signature of DNA nucleotides via transverse transport. Nano Lett 2005;5(3):421−4.

[35] Lagerqvist J, Zwolak M, Di Ventra M. Fast DNA sequencing via transverse electronic transport. Nano Lett 2006;6(4):779−82.

[36] Ivanov AP, Instuli E, McGilvery CM, Baldwin G, McComb DW, Albrecht T, et al. DNA tunneling detector embedded in a nanopore. Nano Lett 2011.

[37] Tsutsui M, Taniguchi M, Yokota K, Kawai T. Identifying single nucleotides by tunnelling current. Nat Nanotechnol 2010;5(4):286−90.

[38] Huang S, He J, Chang S, Zhang P, Liang F, Li S, et al. Identifying single bases in a DNA oligomer with electron tunnelling. Nat Nanotechnol 2010;5(12):868−73.

[39] Tsutsui M, Rahong S, Iizumi Y, Okazaki T, Taniguchi M, Kawai T. Single-molecule sensing electrode embedded in-plane nanopore. Sci Rep 2011;1.

[40] Talaga DS, Li J. Single-molecule protein unfolding in solid state nanopores. J Am Chem Soc 2009;131(26):9287−97.

[41] Han A, Schürmann G, Mondin G, Bitterli RA, Hegelbach NG, de Rooij NF, et al. Sensing protein molecules using nanofabricated pores. Appl Phys Lett 2006;88: 093901.

[42] Fologea D, Ledden B, McNabb DS, Li J. Electrical characterization of protein molecules by a solid-state nanopore. Appl Phys Lett 2007;91(5):053901.

[43] Han A, Creus M, Schürmann G, Linder V, Ward TR, de Rooij NF, et al. Label-free detection of single protein molecules and protein−protein interactions using synthetic nanopores. Anal Chem 2008;80(12):4651−8.

[44] Firnkes M, Pedone D, Knezevic J, Döblinger M, Rant U. Electrically facilitated translocations of proteins through silicon nitride nanopores: conjoint and competitive action of diffusion, electrophoresis, and electroosmosis. Nano Lett 2010;10(6): 2162−7.

[45] Niedzwiecki DJ, Grazul J, Movileanu L. Single-molecule observation of protein adsorption onto an inorganic surface. J Am Chem Soc 2010.

[46] Freedman KJ, Jürgens M, Prabhu AS, Ahn CW, Jemth P, Edel JB, et al. Chemical, thermal, and electric field-induced unfolding of single protein molecules studied using nanopores. Anal Chem 2011.

[47] Yusko EC, Johnson JM, Majd S, Prangkio P, Rollings RC, Li J, et al. Controlling protein translocation through nanopores with bio-inspired fluid walls. Nat Nanotechnol 2011;6(4):253−60.

[48] Kowalczyk SW, Kapinos L, Blosser TR, Magalhães T, van Nies P, Lim RYH, et al. Single-molecule transport across an individual biomimetic nuclear pore complex. Nat Nanotechnol 2011;6(7):433−8.

[49] Wei R, Gatterdam V, Wieneke R, Tampe R, Rant U. Stochastic sensing of proteins with receptor-modified solid-state nanopores. Nat Nano 2012;7(4):257−63.

Index

Note: Page numbers followed by "*f*", "*t*" and "*b*" refer to figures, tables and boxes, respectively.

Printed and bound by CPI Group (UK) Ltd, Croydon, CR0 4YY

08/05/2025

01864838-0008